AXIOMATIC SET THEORY

AXIOMATIC SET THEORY

PAUL BERNAYS

WITH A HISTORICAL INTRODUCTION BY
ABRAHAM A. FRAENKEL

DOVER PUBLICATIONS, INC.
NEW YORK

Published in Canada by General Publishing Company, Ltd., 30 Lesmill Road, Don Mills, Toronto, Ontario.
Published in the United Kingdom by Constable and Company, Ltd., 3 The Lanchesters, 162–164 Fulham Palace Road, London W6 9ER.

This Dover edition, first published in 1991, is an unabridged and unaltered republication of the second edition (1968) of the work first published by the North-Holland Publishing Company, Amsterdam, in 1958 in their series *Studies in Logic and The Foundations of Mathematics.*

Manufactured in the United States of America
Dover Publications, Inc.
31 East 2nd Street
Mineola, N.Y. 11501

Library of Congress Cataloging-in-Publication Data

Bernays, Paul, 1888–
Axiomatic set theory / Paul Bernays ; with a historical introduction by Abraham A. Fraenkel.
p. cm.
Reprint. Originally published: 2nd ed. Amsterdam : North-Holland Pub. Co., 1968.
Includes bibliographical references and index.
ISBN 0-486-66637-9 (pbk.)
1. Axiomatic set theory. I. Title.
QA248.B47 1991
511.3'22—dc20 90-25812
CIP

PREFACE

This monograph is designed for a reader who has some acquaintance with problems of axiomatics and with the standard methods of mathematical logic. No special knowledge of set theory and its axiomatics is presupposed.

The Part of Professor Fraenkel gives an introduction to the original Zermelo–Fraenkel form of set-theoretic axiomatics and an account of its following development.

My part is an independent presentation of a formal system of axiomatic set theory. The formal development is carried out in detail, only in chapt. VII, which is about the applications to usual mathematics, it seemed necessary to restrict myself to some indications of the method of englobing analysis, cardinal arithmetic and abstract algebraic theories in the system. These indications, however, certainly will be sufficient to make appear the possibility of such an englobing.

In composing my part I had the continual and most efficient help of Dr. Gert Müller, with whom I have talked over all details. I express him my very hearty thanks.

To North-Holland Publishing Company and its Director Mr. M. D. Frank I am thankful for the obligingness in the technical questions and the elegant accomplishment of the rather complicated print.

Zürich, March 1958 PAUL BERNAYS

CONTENTS

PREFACE v

PART I. HISTORICAL INTRODUCTION

1. INTRODUCTORY REMARKS 3
2. ZERMELO'S SYSTEM. EQUALITY AND EXTENSIONALITY 5
3. "CONSTRUCTIVE" AXIOMS OF "GENERAL" SET THEORY 9
4. THE AXIOM OF CHOICE 15
5. AXIOMS OF INFINITY AND OF RESTRICTION 21
6. DEVELOPMENT OF SET-THEORY FROM THE AXIOMS OF Z 26
7. REMARKS ON THE AXIOM SYSTEMS OF VON NEUMANN, BERNAYS, GÖDEL 31

PART II. AXIOMATIC SET THEORY

INTRODUCTION 39

CHAPTER I. THE FRAME OF LOGIC AND CLASS THEORY 45
1. Predicate Calculus; Class Terms and Descriptions; Explicit Definitions 45
2. Equality and Extensionality. Application to Descriptions . 52
3. Class Formalism. Class Operations 56
4. Functionality and Mappings 61

CHAPTER II. THE START OF GENERAL SET THEORY 65
1. The Axioms of General Set Theory 65
2. Aussonderungstheorem. Intersection 69
3. Sum Theorem. Theorem of Replacement 72
4. Functional Sets. One-to-one Correspondences 76

CHAPTER III. ORDINALS; NATURAL NUMBERS; FINITE SETS . . . 80
1. Fundaments of the Theory of Ordinals 80
2. Existential Statements on Ordinals. Limit Numbers 86
3. Fundaments of Number Theory 89
4. Iteration. Primitive Recursion 92
5. Finite Sets and Classes 97

CHAPTER IV. TRANSFINITE RECURSION 100
1. The General Recursion Theorem 100
2. The Schema of Transfinite Recursion 104
3. Generated Numeration 109

CHAPTER V. POWER; ORDER; WELLORDER 114
1. Comparison of Powers 114
2. Order and Partial Order 118
3. Wellorder 124

CHAPTER VI. THE COMPLETING AXIOMS 130
1. The Potency Axiom 130
2. The Axiom of Choice 133
3. The Numeration Theorem. First Concepts of Cardinal Arithmetic 138
4. Zorn's Lemma and Related Principles 142
5. Axiom of Infinity. Denumerability 147

CHAPTER VII. ANALYSIS; CARDINAL ARITHMETIC; ABSTRACT THEORIES 155
1. Theory of Real Numbers 155
2. Some Topics of Ordinal Arithmetic 164
3. Cardinal Operations 173
4. Formal Laws on Cardinals 179
5. Abstract Theories 188

CHAPTER VIII. FURTHER STRENGTHENING OF THE AXIOM SYSTEM 195
1. A Strengthening of the Axiom of Choice 195
2. The Fundierungsaxiom 200
3. A one-to-one Correspondence between the Class of Ordinals and the Class of all Sets 203

INDEX OF AUTHORS (PART I) 211

INDEX OF SYMBOLS (PART II) 213
Predicates 213
Functors and Operators 214
Primitive Symbols 215

INDEX OF MATTERS (PART II) 216

LIST OF AXIOMS (PART II) 218

BIBLIOGRAPHY (PART I AND II) 219

PART I

HISTORICAL INTRODUCTION

BY

A. A. FRAENKEL

HISTORICAL INTRODUCTION

1. Introductory remarks

The axiomatic method in mathematics, which started with Euclid's *Elements* and was revived in the 19th century, again chiefly for the purpose of geometry, has made an enormous progress since the beginning of the 20th century; almost all fields of mathematics and logic, and some of physics and other sciences, have since undergone an axiomatic analysis.

While the axiomatic method is appropriate to the homogeneous and continuous domain of geometry to a greater extent than to arithmetic (where a constructive development from simple objects to complicated ones is natural) in set-theory the axiomatic point of view is particularly appropriate for two reasons. First, the antinomies of set-theory which appeared about the turn of the 20th century, show that the quasi-constructive procedure [1]) of Cantor's set-theory has to be restricted in some way, and thus an axiomatic determination of the restriction becomes imperative. Secondly, the fact that all other mathematical branches can be incorporated in set-theory, leads to the idea of setting up a comprehensive axiom system [2]) of set-theory in which the axiomatic theories of other disciplines can be embedded.

It is natural that, after the shock of the antinomies, the stress should be laid on restricting the concept of set axiomatically in such a way that the known contradictions were eliminated and new ones were not to be expected. This was the trend of Zermelo and his followers (since 1908), as described in Nos. 2–6 below. After confidence in the intrinsic soundness of the theory had been re-established by the success of this step, the question arose whether the restric-

---

[1]) A quite different constructive theory, developed by L. E. J. Brouwer since 1907 in accordance with the principles of neo-intuitionism, is outside the subject of the present monograph. See Fraenkel–Bar Hillel [1958], ch. IV.

[2]) A different comprehensive system has been set up in *Principia Mathematica*.

tions imposed on the extent of the set-concept were not exaggerated. Therefore one endeavoured to approach the exact borderline that separates the legitimate theory from the zone of contradictions; this tendency is exhibited in more recent researches (see No. 7 below, and the main part of this monograph [1])).

A few other early attempts to found set-theory axiomatically have either not had sufficient success or not been developed to a point which allows a final judgment [2]).

While a discussion of the axiomatic method in general is beyond the scope of this monograph, a few informal explanations with special reference to set-theory are required. The axiomatization of set-theory renounces a *definition* of the concept of set and of the relation between a set s and its elements. The latter, a dyadic

---

[1]) Cf. Borgers [1949] and the comparative surveys of Zermelo's and other methods in Wang [1949] and [1950] and Wang–McNaughton [1953]. (For all references, see the *Bibliography* at the end of the monograph.)

[2]) Schoenflies [1921] takes the relation between whole and part (proper subset) as the primitive relation. This procedure at best attains a theory of *magnitude* which does not provide for the properties of irreducible parts (elements; see Merzbach [1925]); this also applies to finite sets. — The idea of replacing the element-set relation by the part-whole relation is also the basis of Foradori [1932].

The system of Finsler ([1926], [1933]; Gonseth [1941], pp. 162–180) is based on three axioms only. While by means of the first two axioms it can be proved (see Baer [1928]) that any consistent model of Finsler's system admits of a further extension—as does, for instance, Hilbert's system of geometrical axioms when the axiom of Archimedes is dropped—the third axiom postulates completeness in a sense analogous to Hilbert's axiom. For this reason, as well as for the doubts connected with Finsler's notion *zirkelfrei*, his system is hardly tenable.

The axiomatic system of Gonseth [1933] (cf. [1936]) denies the assumption that, given a set, it is settled whether a given object belongs to the set or not. Hence the fundamental propositions on the non-equivalence of sets forfeit their validity and it is premature to judge the difficulties involved.

The intention of Ting–Ho [1938] is similar to that of Zermelo, but the treatement is not strict enough to allow comparison. Cf. also Giorgi [1941]. Systems of a "logicistic" type, such as Ramsey [1926], Quine [1937] and [1940] (cf. [1941] and [1942]), Wang [1954], are not included in the subject-matter of this monograph; neither Lorenzen [1955].

relation (or predicate), is denoted by ε; $x \varepsilon s$ reads "x is contained in, is an element of, belongs to, the set s" or "s contains (the element) x", and its negation is $x \bar{\varepsilon} s$. ε enters as an (undefined) *primitive* relation, *the membership relation.* It is unsymmetrical and the values of its second argument, possibly with the addition of the null-set (see below), constitute the domain of *sets.* Certain statements containing the membership relation and relations defined through it will be introduced as *axioms.* A statement is *true* if and only if it can be deduced from the axioms (by means of a suitable system of logic, in particular certain rules of inference), and the same applies to the *existence* of sets.

The situation is still simpler in the modification **Z** of Zermelo's system given in Nos. 2–6 below; here – as opposed to Zermelo's own system – no other objects than sets appear, hence the first and the second arguments of the membership relation determine the same domain. On the other hand, in the systems briefly mentioned in No. 7 and extensively treated in the main part of this monograph, the second arguments of the membership relation may belong to a different domain, the domain of *classes.*

As to the elimination of contradictions, all that can be expected is the exclusion of the logical and semantical antinomies known at present; this is attained by the exclusion of "overcomprehensive" sets in the system **Z** (cf. No. 7) and by the formulation of Axiom **V** in No. 3. Sufficient hints at how the essential parts of classical set-theory can be derived from the axioms of Nos. 2–5 are given in No. 6. A few remarks only about the independence of axioms occur in this historical part; the question is discussed in a more profound way in the main part of the present monograph.

2. Zermelo's system [1]). Equality and extensionality

The introduction to the main part of this monograph, written by Bernays, begins with a reference to the historically first axiomatization of set-theory, given by Zermelo in 1908. *The chief*

[1]) The main source is Zermelo [1908a]; see also [1930]. Cf. the expositions in Fraenkel [1927] and [1928] (also Cavaillès [1938], Weyl [1946]), Ackermann [1937], Church [1942] (pp. 180–181).

purpose of this part of the monograph is to give a historical introduction through an exposition of Zermelo's system, with some improvements which were inserted into it before the fundamental changes performed by von Neumann and Bernays from 1925 and 1937 on (see No. 7).

Prior to a systematic exposition, we start with some informal remarks to motivate the method adopted afterwards.

Within a certain non-empty domain of objects we take, as the only primitive relation of our axiomatic system **Z**, the membership relation ε (see above). If x and y denote any objects of the domain, the statement $x \varepsilon y$ shall either hold true or not. While those parts of logic that are necessary for **Z**, in particular the rules of inference and quantification with respect to thing-variables, are assumed to be pre-established, the relation of *equality* should be treated explicitly. Here the following attitudes are possible [1]).

1) Equality in its *logical meaning as identity*. Zermelo adopts this attitude by calling x and y equal "if they denote the same thing (object)". When the objects are sets he in addition rests on an axiom ot extensionality (see below) which states that a set is determined by its elements.

Thus Zermelo's axiom differs intrinsically from the Axiom of Extensionality as expressed below (p. 8), which refers to *all* objects of the domain [2]).

2) Equality as a (second) *primitive relation* within **Z**. Then the usual properties of any equivalence (equality) relation must be guaranteed axiomatically, in particular substitutivity with regard to ε in the two-fold sense: of extensionality as above, and of equal objects being elements of the same sets.

3) Equality as a *mathematically defined relation*. We may define $x=y$ either by "if every set that contains x contains also y and vice versa", or by "if x and y contain the same elements". The

[1]) Cf. Fraenkel [1927] and [1927a], A. Robinso(h)n [1939]. For a more general attitude, cf. Hailperin [1954]. See also the main part of the present monograph.

[2]) The situation becomes somewhat different if, as done in Quine [1940], every "individual" (p. 7) is regarded as a unit-set containing itself.

second way is possible only if, as assumed in the following, every object is a set (including the null-set). In the former case, extensionality must be postulated axiomatically; in the latter, an axiom has to guarantee the former property.

In this part we adopt method 3), which seems superior to 2) insofar as a single primitive relation only occurs in the system, and to 1) since the system is constructed upon a weaker basic discipline. It makes no essential difference which of the two definitions of equality is chosen, provided we take a suitable decision about the existence of *individuals* (called *Urelemente* in Zermelo [1930]), i.e. of objects which contain no element [1]).

Taking into account the admissibility of a set which has no element ("null-set"), three positions about individuals are tenable: that the domain contains one null-set and also other individuals, individuals but no null-set, one null-set but no other individuals. (A domain without null-set and individuals would be impractical.) The first position was taken by Zermelo and, for instance, by Ackermann [1937a]; the second by Quine from 1936 on; the third, first proposed in Fraenkel [1921/22] and later accepted by von Neumann, Bernays, and others, is adopted in the following. This involves that all objects of **Z** are sets, hence that the values of the first and of the second argument of the membership relation constitute the same domain. In fact, for the purpose of developing mathematics it has proved unnecessary to assume the existence of individuals. However, as pointed out below in No. 4, there are problems of independence for which the assumption that infinitely many individuals can be admitted to the domain plays an important part; thus it appears that those problems are more difficult within the system **Z** than in Zermelo's original system.

We now outline the system **Z**, which is not empty and whose only primitive relation is the dyadic relation ε of membership whose arguments are sets.

---

[1]) This use of "individual" has nothing to do with the distinction between "individuals" and "classes" in logic (cf., for instance, Tarski [1935], § 2, and the main part of this monograph). In the logical sense the sets are individuals.

Definition I. If s and t are sets such that, for all x, $x \varepsilon s$ implies $x \varepsilon t$, s is called a *subset* of t, in symbols $s \subseteq t$; in particular a *proper subset* ($s \subset t$) if there is a $y \varepsilon t$ with $y \bar{\varepsilon} s$.

In contrast with Cantor's "comprehensive" method of constructing sets, this definition does not allow the construction of subsets of t by "collecting" some of its elements. Only with regard to given sets may we state that one is a subset of the other.

It follows that the relation $\subseteq$ is reflexive ($s \subseteq s$) and transitive (i.e., $s \subseteq t$ and $t \subseteq u$ imply $s \subseteq u$); $\subset$ is irreflexive, transitive, and asymmetrical (i.e., $s \subset t$ and $t \subset s$ are incompatible).

In accordance with earlier remarks, *equality* is defined in either of the following ways.

Definition IIa. If, for all x, $s \varepsilon x$ implies $t \varepsilon x$ and conversely, s equals t ($s = t$); the negation is $s \neq t$ (s differs from t). That is to say, sets are equal if contained in the same objects (sets).

Definition IIb. If $s \subseteq t$ and $t \subseteq s$, then $s = t$; otherwise $s \neq t$. That is to say, sets containing the same objects are equal.

Equality is a reflexive, symmetrical, and transitive relation. The definitions are somehow peculiar to **Z**; in fact, in the systems of No. 7 (below) not every object can become an element of another object, as against IIa, while in Zermelo's own system there may exist different objects without elements, as against IIb.

Equality is substitutive with regard to the second argument of ε, i.e. from $x \varepsilon s$ and $s = t$ it follows that $x \varepsilon t$ [1]). But IIa does not yield extensionality nor does IIb yield substitutivity regarding the first argument; hence we supplement IIa and IIb respectively with the axioms

Axiom **Ia.** $s \subseteq t$ and $t \subseteq s$ imply $s = t$.

Axiom **Ib.** $x \varepsilon s$ and $x = y$ imply $y \varepsilon s$.

It makes no difference whether we adopt Definition IIa and Axiom **Ia**, or IIb and **Ib**. Hence we shall simply speak of *the Definition* (II) *of Equality* and of *the Axiom* (**I**) *of Extensionality.*

---

1) This is evident in view of IIb; for the proof in view of IIa, cf. A. Robinso(h)n [1939], footnote 4.

Since a set is determined [1]) by its elements, we denote the set with the elements $a, b, c, \ldots$ also by $\{a, b, c, \ldots\}$, regardless of the order of the elements.

Definition III. Two sets without common elements are called *mutually exclusive*. If s contains at least two elements, and any two elements of s are mutually exclusive, s is called a *disjointed* set.

3. "Constructive" axioms of "general" set theory

In the heading two *ad hoc* terms are used. "Constructive" means that, certain things (one set, two sets, a set and a predicate) being given, the axiom states the existence of a *uniquely determined* other set; "general" theory means – in contrast with the use in the main part of the monograph; cf. Bernays [1942a], p 133 – that no axioms of infinity (see No. 5) are included. (According to the present attitude the Axiom of Power-Set, for instance, is not an axiom of infinity.)

The axioms will be preceded by informal remarks which point out the immediate purpose of every single axiom and, thereby, hint at their independence.

The operation of "uniting" two different sets [2]) is introduced by

> *Axiom* II *of Pairing* [3]). For any two different sets a and b, the pair $\{a, b\}$, or $\{b, a\}$, exists.

On account of extensionality we are entitled to use the definite article (*the* pair) here and in the following three axioms, as well as in Theorems 1–3 below.

---

[1]) Hence Zermelo's name for Axiom **Ia**: *Axiom der Bestimmtheit*. Yet he explicitly restricts the axiom to the case that s and t contain elements, a restriction not adopted in our system **Z**.

[2]) Instead of this Zermelian operation, Kuratowski [1925] uses the union of two sets in the sense of Axiom **III**. — For the case $a = b$, see theorem 1 on p. 14.

[3]) In Zermelo's terminology, *Axiom der Elementarmengen*. This includes postulating the null-set and the set containing a single given element; both will be *proved* to exist in the present exposition.

Given any number of sets, Axioms **I** and **II** do not enable us to produce new sets other than pairs. The primary operations of Boolean algebra, union and intersection, suggest themselves as simplest additional procedures, and the former will prove sufficient. We introduce it by

> *Axiom* **III** *of Sum-Set (Union).* For any set s which contains at least two elements, there exists the set whose elements are the elements of the elements of s.

This set is called the *sum-set* of s, or the *union of the elements* of s, and is denoted by $\bigcup s$. If $s=\{a, b\}$ we also write $a \cup b$ for $\bigcup s$, and if s contains the elements $a, b, c, \ldots$ we write $a \cup b \cup c \cup \ldots$ for $\bigcup s$.

The union of two different sets exists by **II** and **III**. Given a number of sets, certain types of new sets can be produced, i.e. proved to exist, and the associativity of the union-operation is easily shown. Nevertheless, clearly these axioms do not enable us to proceed to more-than-denumerable sets if, say, a sequence of denumerable sets is given, where "denumerable" and "sequence" are informal terms to be formally defined later.

Cantor's (second) tool for reaching higher powers was the operation of (transfinite) multiplication, in particular exponentiation. We can, however, content ourselves with the special tool of the power-set, i.e.

> *Axiom* **IV** *of Power-Set.* For any set s, there exists the set whose elements are all subsets of s.

This set is called the *power-set* of s and denoted by $\Pi(s)$ [1]).

The efficacy of this axiom differs from that of Cantor's respective operation not only in that the existence of s is presupposed, but in that the existence of the subsets of s is here assumed to be previously established. Since Axioms **I–III** yield only few very special subsets of a given set and since Definition I does not

[1]) Zermelo writes $\mathfrak{S}s$ for $\bigcup s$ (Axiom **III**), and $\mathfrak{U}s$ for $\Pi(s)$. Thus also, for instance, in Kleene [1952]. In the main part of this monograph $\sum s$ is used for $\bigcup s$.

enable us to produce any subsets [1]), Axiom **IV** is at the present juncture a very limited instrument and not at all sufficient to yield the so-called "theorem of Cantor" about the cardinality of the power-set. For instance, the existence of *infinite* proper subsets of an infinite set s cannot be ensured so far. Hence methods of producing subsets of a given set remain the chief desideratum – and, as we shall see *a posteriori*, the only one left within general set theory. The principal method in **Z** (for an additional direction see No. 4) is given by

> *Axiom* **V** *of Subsets* [2]). For any set s and any predicate $\mathfrak{P}$ [3]) which is meaningful ("definite") for all elements of s, there exists the set y that contains just those elements x of s which satisfy the predicate $\mathfrak{P}$ (the condition $\mathfrak{P}(x)$).

y is clearly a subset of s.

The weak point in this formulation of the axiom is the term "meaningful predicate" (or property); in Zermelo's terminology, *definite Eigenschaft.*

Informally this term may be understood to mean that, for each $x \varepsilon s$, $\mathfrak{P}(x)$ should be either true or false, without demanding that the decision ought to be reached at the present stage of scientific development. Thus "x is transcendental" is meaningful when s is a set of numbers, but not "x is finitely definable" or another semantic condition, as those appearing in the antinomies of the semantical type.

Clearly, such explanations cannot satisfy the requirements of a formal deductive theory. Zermelo [1908a] (p. 263) gave the following paraphrase: *Eine Frage oder Aussage* $\mathfrak{E}$, *über deren Gültigkeit oder Ungültigkeit die Grundbeziehungen des Bereiches* [4])

---

[1]) This is also the case in Zermelo's exposition, but has been misunderstood by some of his interpreters.

[2]) Zermelo calls it *Axiom der Aussonderung* (of "sifting").

[3]) $\mathfrak{P}(x)$ is what is called by Rosser [1953] (e.g., p. 200) *a condition on* x; other expressions are "a statement with free occurrence of x" or "a well-formed formula".

[4]) The intention is to the membership relation, and presumably also to the equality relation.

vermöge der Axiome und der allgemeingültigen logischen Gesetze ohne Willkür entscheiden, heisst "definit". Ebenso wird auch eine "Klassenaussage" 𝔈(x), in welcher der variable Term x alle Individuen einer Klasse 𝔎 durchlaufen kann, als "definit" bezeichnet, wenn sie für jedes einzelne Individuum x der Klasse 𝔎 definit ist. So ist die Frage, ob $a \varepsilon b$ oder nicht ist, immer definit, ebenso die Frage, ob $M \subseteq N$ oder nicht.

13 years then elapsed until the first steps were taken to replace this hardly satisfactory explanation by a more rigorous one. Fraenkel [1921/22] and Skolem [1922/23] independently took two seemingly different directions which, however, proved to be essentially equivalent [1]), and of which the second is preferable for its more general and natural character.

The first method formalizes "definiteness" by means of a special concept of function defined by the operations of Axioms **II–V**; the inclusion of **V** itself leads to a certain hierarchy of orders, in accordance with the fact that the axiom constitutes an axiom schema. The actual derivation of classical (general) set-theory, comprising the theories of order and well-order, shows that this apparently special concept of function is sufficient [2]).

The second method [3]) formalizes "definiteness" by using the concept of (elementary) formula, i.e. of "statement" in the sense of Rosser [1953] (p. 208), obtained from variables and the membership relation by negation, conjunction, disjunction, and quantification with respect to thing-variables, within the first-order predicate calculus with its truth-functions. As proven by Skolem, this procedure covers the first method; it, too, shows the axiom to represent an axiom schema which contains infinitely many particular axioms.

[1]) Cf., in particular, Skolem [1929]. For the first method, cf. Fraenkel [1922] and [1927] and von Neumann [1928a].

[2]) See below No. 6. For existence theorems as those connected with ordinal numbers, the Axiom of Substitution (No. 5), with the generalized function concept of von Neumann [1928a], is required. Cf. also Curry [1934], p. 590, and [1936], p. 375.

[3]) Cf. Skolem [1929] (§ 2) and [1930]; there Schröder's *Algebra der Logik* is used, but this is not an essential feature.

On the other hand, Quine [1]) suggests an extension of the Axiom of Subsets which would render Axioms **II–IV** (above) unnecessary.

Zermelo, however, rejected Fraenkel–Skolem's methods of formalizing his concept *definite*, particularly because they implicitly involve the concept of natural number which, in Zermelo's view, should be based upon set theory. Instead he introduced, in [1929], a special and rather complicated axiomatization of the concept of definiteness (or of function) imbedded in his axiom system [1908a]. This method suffers from various disadvantages pointed out in Skolem [1930]; to be sure, through the increase in primitive concepts the axiom of subsets is transformed from an axiom schema into an axiom proper.

In what follows, no explicit attitude among those mentioned regarding "definiteness" in Axiom **V** will be adopted. Yet implicitly we shall be guided by Skolem's conception.

On account of the Axiom of Subsets, the Axiom of Power-Set gains its actual strength. The connection between these two axioms is essentially impredicative; whenever, for instance, the predicate used in Axiom **V** refers to the power-set of s, a particular subset of s is defined by means of the totality of all its subsets. In fact, we have to consider Axiom **V**, in contrast with the preceding axioms, to be an *axiom schema* [2]) which yields infinitely many single axioms corresponding to the predicates chosen, and the system **Z** proves not to be *finitizable* [3]), i.e. reducible to finitely many axioms proper – in contrast with the systems appearing below in No. 7, or to Quine [1937] (or [1953]) [4]); on the other hand, in the latter systems the number of primitive concepts is greater than in **Z**, or else the "metalogical" rule corresponding to Axiom **V** is added to the usual rules.

---

[1]) Quine [1936], [1937], [1953]; in this system no null-set appears. A similar attitude is taken in Lindenbaum–Mostowski [1938] and other researches of these authors.

[2]) This concept was first introduced in von Neumann [1927], p. 13.

[3]) See Wang [1952]; consult, however, Wang [1955], pp. 82/3. Cf. Mostowski [1951] and [1953].

[4]) The proof in Hailperin [1944] is rather surprising since Quine's original class axiom schema is impredicative.

Finally, we draw a few conclusions from Axioms **I–V**.

Definition. A set which contains no element is called a *null-set*.

By extensionality, there exists at most one null-set, which shall be denoted by O. It clearly is a subset of every set.

Theorem 1. There exists one null-set and, for every set t, the unit-set $\{t\}$ whose only element is t.

Proof. First, for a set $s \neq O$ of the non-empty system **Z**, we take a predicate which is not satisfied by any element x of s (e.g., $x \bar{\varepsilon} s$), and obtain O by Axiom **V**. Secondly, if $t \neq O$, we obtain $s = \{t, O\}$ by Axiom **II**; then the predicate $x = t$ (or $x \neq O$) gives, by Axiom **V**, the subset $\{t\}$. If $t = O$, $\{O\}$ exists by Axiom **IV**.

Theorem 2. For every non-empty set t, there exists the set whose elements are common to all elements of t.

It is called the *intersection* of the elements of t and denoted by $\bigcap t$. If $a, b, \ldots$ are the elements of t, the intersection is also denoted by $a \cap b \cap \ldots$.

Proof. By Axiom **III**, $s = \bigcup t$ exists and contains the elements of the elements of t. The predicate "x is contained in each element of t" defines, by Axiom **V**, a subset of s which is the intersection $\bigcap t$.

Theorem 3. For every disjointed set t, there exists the set whose elements are the sets which contain a single element from each element of t.

It is called the *Cartesian product* or cross product (in Cantor's terminology, Verbindungsmenge) of the elements of t, and shall be denoted by $\mathfrak{P}t$. If $a, b, \ldots$ are the elements of t, the Cartesian product is also denoted by $a \times b \times \ldots$. Clearly $\mathfrak{P}t = O$ if $O \varepsilon t$.

Proof. Every set that contains a single element from each element of t is at any rate a subset of $\bigcup t$. By Axioms **III** and **IV**, there exists the power-set $s = \Pi(\bigcup t)$. We take the predicate "$x \varepsilon s$, and for each $\tau \varepsilon t$ the intersection $\tau \cap x$ is a unit-set"; hence, by Axiom **V**, we obtain the Cartesian product $\mathfrak{P}t$.

4. The axiom of choice [1])

After admitting the Axiom of Subsets, the question still remains open whether other subsets of a given set, not uniquely characterized by a certain predicate, are conceivable or even required. To answer this question (at least, partly) we proceed as follows.

Let t be a disjointed set with $O \bar{\varepsilon} t$. According to Theorem 3 on p. 14, the elements of the Cartesian product $\mathfrak{P}t$ are those subsets of $\bigcup t$ whose intersections with each element of t are unit-sets. Yet the proof of the theorem does not show whether the assumption $O \bar{\varepsilon} t$ implies that $\mathfrak{P}t \neq O$.

True, our assumption seems to suggest the possibility of "choosing" a single arbitrary element in each $\tau \varepsilon t$; the set c that contains just the elements chosen would be a subset of $\bigcup t$ with the property desired, hence an element of $\mathfrak{P}t$, which shows that $\mathfrak{P}t \neq O$. Yet except for the trivial case where each element of t is a unit-set, our argumentation does not show how the set c can be obtained by means of Axiom **V**. On the other hand, *if* such a set c exists, additional subsets of $\bigcup t$ with the property desired can be easily constructed from c, for example by replacing a definite element of c with another from the same element of t.

We therefore introduce a special axiom, the last one required for "general" set theory, namely

> *Axiom* **VI** *of Choice* (or Multiplicative Axiom [2])). For every disjointed set t for which $O \bar{\varepsilon} t$, the Cartesian product $\mathfrak{P}t$ differs from the null-set.

Each element of $\mathfrak{P}t$ shall be called a *selection-set* of t, as it contains a single element from each element of t.

In contrast with Axioms **II**–**V**, a set produced by Axiom **VI** is *not uniquely determined* by its data, i.e. by the set t. Hence **VI** is

---

[1]) For additional material to this subject, see the main part of this monograph, and Chapter II, § 4, of Fraenkel–BarHillel [1958].

[2]) This is the (appropriate) name due to B. Russell; in fact, the axiom guarantees that a product of cardinals $\neq 0$ is itself $\neq 0$. "Axiom of choice" (*Auswahlaxiom*, *axiome du choix*, etc.) is the name given by Zermelo [1904] for obvious psychological reasons; though less suitable, this name has by now been universally accepted.

not "constructive" in the sense of p. 9; in fact, misunderstandings with regard to the purely existential character of the statement of Axiom **VI** have led to many sterile discussions during some decades [1]).

Axiom **VI** was formulated as above by Russell [1906]. Zermelo [2]) formulated it for the general case where t need not be disjointed and, accordingly, the "chosen" elements may be the same for different elements of t. Then, using the concept of a function which can be defined in **Z** (cf. No. 6), we obtain the following *generalized* principle:

> For every non-empty set t for which $O \bar{\varepsilon} t$, there exists at least one single-valued function $f(\tau)$ whose argument τ runs over the elements of t, such that $f(\tau) \varepsilon \tau$ [3]).

As shown in Zermelo [1908a], this generalized form can be derived from Axiom **VI** by means of the Axioms **I–V** [4]).

The Axiom of Choice (together with the continuum-hypothesis, cf. p. 25) is probably the most interesting and most discussed axiom in mathematics after Euclid's axiom of parallels.

Fundamental problems regarding the Axiom of Choice are, a) is it *independent* of Axioms **I–V** (and possibly of the axioms of No. **5**)? b) is it *compatible* with these axioms? The latter question was answered in the affirmative by Gödel [1938] and [1940], both within a modification of Bernays' axiom system (cf. No. **7**) and within other systems such as **Z** or a modified form of *Principia Mathematica*. However – notwithstanding partial results specified below – it is unknown as yet whether the axiom is independent or whether it can be proved within the system **Z** or similar systems.

---

1) Cf. the (justified) attitude of Ramsey [1926], and also the position of the statement of **VI** in von Neumann's system.

2) Zermelo [1908a]; it was used earlier, in particular cases, in Zermelo [1904] and [1908]. The idea occurs already in B. Levi [1902]. The existence of selection-sets had been used before, inadvertently and without a proper argument, by Cantor and others.

3) A further generalization is given in Skolem [1929]. Cf. the main part of the present monograph.

4) Cf. below p. 28, also the final chapter of Church [1944].

In particular, one has not succeeded in solving this problem, for instance, for sets t whose elements are *arbitrary sets of real numbers.*[1]) (Independence would mean that no selection-set of t can be provided by means of the Axiom of Subsets.) Nevertheless, and though we do not even know in what direction an ultimate independence proof might be looked for [2]), there are strong indications of the independence of the axiom, notably the wide use of the axiom in important proofs without a visible alternative and, on the other hand, the proofs of independence reached under weaker assumptions; for instance, within Zermelo's original system (see below).

The most obvious *specializations* [3]) of Axiom **VI** are obtained by adding particular assumptions about the *cardinality of the set t and/or of its elements.*

If t is finite the axiom becomes redundant, viz. provable by means of the other axioms; it then expresses a distributive law connecting logical conjunction and disjunction. In fact, if t contains a single non-empty set, the existence of a selection-set follows from the existence of the unit-set (Theorem 1, p. 14) by some simple steps in predicate calculus [4]); this result is transferred to any finite t by mathematical induction.

On the other hand, *the finiteness of the elements of t* does not trivialize the existence of a selection-set; hence the *weakest form* of the axiom refers to the case where every element of t is a pair or, more generally, a finite set with a cardinal > 1 [5]).| Furthermore,

[1]) Already Lebesgue [1907] had shown that in such cases no "analytically representable" function can define a selection-set. In Gonseth [1941] (p. 117) Lebesgue showed by a concrete example that the distinction between construction and existence may, through the Axiom of Choice, affect elementary (geometrical) problems.

[2]) Whether Gandy [1956] is apt to yield such a proof remains to be seen.

[3]) Specializations restricted to the theory of sets of points are Beppo Levi's *principio di approssimazione* (Levi [1923] and [1934]) and Knaster's hypothesis referring to linear perfect sets (cf. Kondô [1937]).

[4]) Erroneously several distinguished scholars, in particular A. Denjoy, maintained that $t=\{\dot{\tau}\}$ $(\tau \neq O)$ was the crucial case of the axiom of choice, form which the general case might be easily inferred.

[5]) Fraenkel [1922].

if t is denumerable, we speak of the *restricted axiom of choice* [1]), no matter what the cardinals of the elements of t may be.

Regarding the *independence of the Axiom of Choice* in its various forms, i.e. the impossibility of proving its statement within a suitable system, quite a number of interesting and profound results have been obtained *under the assumption that the system may contain infinitely many, or even more-than-denumerably many, different objects which contain no element*, called *Urelemente* by Zermelo. This assumption, which is necessary for the construction of suitable models showing the independence, is incompatible with the axioms of our system **Z** because of the Axiom of Extensionality (nor is it consistent within the systems described below in No. 7); hence, as stated before, in such systems the independence of the axiom is an open problem. The assumption is consistent in Zermelo [1908] and [1930]; however, it seems neither useful nor justified from a general point of view but is an *ad hoc* resource [2]).

Quite recently the use of *Urelemente* in these independence proofs has been replaced by the use of *ensembles extraordinaires* (p. 24) [3]). While this in a certain sense is a progress, the main problem (e.g., for a set of sets of real numbers) remains unsolved as before.

The most important independence problems in question are the following:

1) Can the weakest form of the axiom be proved, at least for denumerable t?

2) What interdependence exists between various kinds of the weakest form with respect to the cardinals of the finite sets which are the elements of t?

3) Can every infinite set be ordered within the system? (Order can be defined by means of the membership relation, see No. 6.)

[1]) Tarski [1948]. In the same paper and before in Bernays [1942] (p. 86, axiom IV*), more closely in Mostowski [1948], another weakened form is considered, the "principle of dependent choices" (see below p. 19).

[2]) It is characteristic that this assumption is also required for Essenin–Volpin's [1954] proof that Souslin's problem cannot be answered in the affirmative without the Axiom of Choice.

[3]) Mendelson [1956a]. Cf. Bernays [1954] (p. 84), Shoenfield [1955], Specker [1957].

4) Does the assumption that every infinite set can be ordered, imply the well-ordering theorem (which is known to be equipollent to the Axiom of Choice)?

5) Does the assumption of the weakest form imply the general statement of the axiom? (This question admits of various cases according to the cardinality of t.)

6) Can the general statement be proved by means of the following *axiom of dependent choices* (which certainly implies the restricted form, while the converse still seems to be an open question): if B is a non-empty set and ϱ a binary relation such that for every $x \varepsilon B$ there is a $y \varepsilon B$ with $x \varrho y$, then there exists a sequence $(x_1, x_2, \ldots, x_k, \ldots)$ of elements of B such that $x_k \varrho x_{k+1}$ for every k.

Negative answers [1]) to the problems 1), and 3) to 6), have been obtained only within systems in which the assumption of p. 18 can be fulfilled, while problem 2) is investigated to a great extent in Mostowski [1945] and Szmielew [1947], both in a group-theoretic and a number-theoretic direction. Some well-known consequences of the Axiom of Choice, for instance that every infinite (i.e., non-inductive) set includes a denumerably infinite subset, or that the sum-set of every disjointed infinite set t includes a subset which is equivalent to t, were proved to be independent in the same sense and with similar methods by Lindenbaum and Mostowski.

Finally, there is the problem of classifying the various definitions of finiteness into different categories such that the transition from the definitions of one category to those of another involves the Axiom of Choice in one but not in the converse direction. A review of particular cases of this problem was already given in the *Annexe* of Tarski [1925] and quite a number of them were solved by Lindenbaum and Mostowski (see the footnote) in a negative sense as mentioned.

---

[1]) Fraenkel [1922], [1928a], [1937] (see, however, Lindenbaum–Mostowski [1938]), Mostowski [1938], [1939], [1948]. Fraenkel's group-theoretic method of proof, while adopted by Mostowski and Lindenbaum, was based by them upon a stricter logico-mathematical foundation and supplemented with the method of "relativization of quantifiers" in the sense of Tarski [1935a] and Lindenbaum–Tarski [1936].

The numerous applications of the Axiom of Choice in most branches of mathematics, particularly in analysis, topology, and set theory [1]), are well-known [2]). In abstract set theory the most important statements equipollent to the axiom are the well-ordering theorem (Zermelo [1904] and [1908]) and the comparability of cardinal numbers (Hartogs [1915]). In arithmetic the axiom appears in connection with the notion of finite set or number (cf. above). The application of the axiom in algebra, particularly abstract algebra, dates from Steinitz [1909]; however, for the last twenty years the use of the axiom in algebra (through the well-ordering theorem and transfinite induction) has more and more been replaced by an equipollent *maximum principle*, either in the form of Zorn's lemma (Zorn [1935]) or otherwise. These principles, which originate from Hausdorff [1914] (pp. 140 f.) and have been rediscovered several times independently, are discussed in detail in the main part of the present monograph; cf. also, for instance, Birkhoff [1948].

Finally it should be stressed that to-day, after more than fifty years, there is still a lot of controversy regarding the Axiom of Choice [3]), not only among mathematicians in general but in particular among those working in set theory. (Intuitionists, of course, reject the axiom because of its purely existential character – except for Poincaré who was ready to admit it.) The chief controversial points up to 1938 may be gathered from Zermelo [1908], Borel [1914] (note IV), and Gonseth [1941]. Some authors, e.g. Borel and Denjoy, think the axiom rather acceptable if t is a denumerable set, while the scepticism of Lebegue and others makes no distinction between denumerable and non-denumerable t. The opposition to the axiom is partly based on certain

[1]) For the connection between the axiom and the generalized continuum-hypothesis, we refer to Specker [1954]; cf. already Lindenbaum–Tarski [1926].

[2]) The earliest survey is given in Sierpiński [1919].

[3]) Hilbert [1923] maintains *dass der wesentliche dem Auswahlprinzip zugrunde liegende Gedanke ein allgemeines logisches Prinzip ist, das schon für die ersten Anfangsgründe des mathematischen Schliessens notwendig und unentbehrlich ist.*

paradoxes following from it, such as Banach–Tarski's theorem. There exist also inquiries into the possible consequences of assertions *negating* the Axiom of Choice [1]), in some external analogy to non-Euclidean geometries.

5. Axioms of infinity and of restriction

Set-theory as derived from Axioms **I–VI** does not enable us to prove the existence of an *infinite set.* (On the other hand, as long as finite sets alone are considered, part of the axioms are redundant, notably **V** and **VI**, and essentially also **IV**.) Bolzano's and Dedekind's alleged proofs of the existence of infinite sets are known to be illusory. Therefore Zermelo [1908a] introduced the following [2])

> *Axiom* **VII** *of Infinity.* There exists at least one set W with the properties
> a) $O \varepsilon W$
> b) if $x \varepsilon W$, then $\{x\} \varepsilon W$.

If W is an arbitrary set with these properties and U the set of those subsets of W which have the same properties, clearly the intersection $\bigcap U = W_0$ also has them. W_0 proves independent of the particular set W; it is the least set that contains the elements O, $\{O\}$, $\{\{O\}\}$, ... and may be considered to be the set of non-negative integers. In particular, W_0 is an "infinite" set, viz. equivalent (cf. No. 6) to a proper subset in the sense that the mapping required can be constructed in **Z**.

More extensive infinite sets can then be obtained by means of the axioms of No. 3, in particular the Axiom of Power-Set.

---

1) Church [1927] investigates the influence of such negation upon the theory of the second number-class. For various alternatives to the axiom, cf. Bachmann [1955], § 39, and Specker [1957].

2) A more general, though equipollent, formulation is given in Wang [1949], p. 151. A far stronger form, introduced in Fraenkel [1927] (p. 114, Axiom VIIc), is utilized by R. M. Robinson [1937] (Axiom 8.3) and Bernays [1942] (Axiom VI*). This stronger form where, instead of $\{x\}$, more general functions of x are admitted, proves independent of the system of Axioms I–VII (Rosser–Wang [1950], p. 128).

A similar way of ensuring the existence of infinite sets is the following axiom [1]) which, though less simple in its formulation, yields a more lucid and useful alternative to W_0.

Axiom **VII***. There exists at least one set W^* with the properties

a*) $O \varepsilon W^*$

b*) if $x \varepsilon W^*$, then $(x \cup \{x\}) \varepsilon W^*$.

In a way corresponding to that used above, we infer the existence of a minimal set W_0^* with the properties a*) and b*), which contains the elements

$$O, \{O\}, \{O, \{O\}\}, \{O, \{O\}, \{O, \{O\}\}\}, \ldots$$

The set W_0^* can as well be considered to be the infinite set of non-negative integers. Of any two different elements of W_0^*, one is both an element and a proper subset of the other, and hereby a natural order is established in W_0^*.

In 1921 it turned out that Axioms **I–VII**, which approximately correspond to Zermelo's original seven axioms, were not sufficient for the legitimate needs of set theory; the simplest instance of a set which cannot be obtained by those axioms is a set P which, in addition to an arbitrary infinite set (e.g., W_0), contains the power-set of each set contained in P. To fill this gap in a way which yields particular sets such as just mentioned as well as the general theory of ordinal numbers and of transfinite induction, an axiom of the following type is required: [2])

Axiom **VIII** *of Substitution* (or Replacement). For every set s and every single-valued function f of one argument which is defined for the elements of s, there exists the set that contains all $f(x)$ with $x \varepsilon s$.

In short, if the domain of a single-valued function is a set, its counter-domain is also a set.

---

[1]) It conceives the integers as ordinals in the sense of von Neumann [1923].

[2]) Fraenkel [1921/22] and [1927], Skolem [1922/23] and [1929], von Neumann [1923] and [1928a]; cf. Lindenbaum–Tarski [1926]. For the independence of **VIII** of the system **I–VII**, cf. the literature given below on p. 35.

An alternative formulation [1]), preferable in some respects, reads: If s is a set and φ a binary relation between sets such that for every $x \varepsilon s$ there is just one y satisfying $\varphi(x, y)$, then there is a set t such that $y \varepsilon t$ is equipollent to the existence of an $x \varepsilon s$ that satisfies $\varphi(x, y)$.

The function-concept entering into **VIII** is closely related to the predicate concept of Axiom **V**. Like **V**, also **VIII** is not a single axiom but an axiom schema.

VIII yields both the stronger form of Axiom **VII** (footnote 2 on p. 21) and the Axiom of Subsets (**V**). Yet it would not be practical to drop **V**, for **I–VII** are sufficient for the bulk of set theory while **VIII** is required for certain purposes only.

Whereas Axiom **VII** and **VIII** *extend* our system by warranting comprehensive sets of certain types, some way of *restricting* the system is desirable as well. It is more than doubtful whether this purpose can be reached by a *general* postulate [2]) – a kind of counterpart of Hilbert's *Vollständigkeitsaxiom* – which restricts the system to the minimal extent compatible with **I–VIII**, similar to the task of Peano's axiom of mathematical induction with respect to his other axioms.

However, there are a few problems which suggest the introduction of more particular restrictive axioms. The chief problems are (a) inaccessible numbers, (b) "extraordinary" sets, (c) the (generalized) continuum problem.

The first problem concerns the existence of regular initial (ordinal) numbers ω_α with a limit-index α, and the corresponding cardinals, called *inaccessible numbers*. Within **Z** the question reads whether *sets* with such ordinals or cardinals exist. The question was broached by Hausdorff in 1908 (see, in particular, Hausdorff [1914], p. 131) and by Mahlo in 1911 while more profound investigations started with Sierpiński–Tarski [1930] and, particularly,

---

[1]) Cf. Skolem [1929], Church [1942] (p. 181). A weakened form of the axiom is given in Tarski [1948], p. 80.

[2]) Suggested in Fraenkel [1921/22], rejected in von Neumann [1925], vindicated in Suszko [1951] and Carnap [1954] (p. 154).

Tarski [1938]. Moreover, inaccessible numbers meanwhile turned out to have significance for certain concrete problems (for instance, of the theory of measure) and not only for the foundations of set theory. For manifold recent investigations on inaccessible numbers and, particularly, for the distinction between "inaccessible" in a narrower [1]) and a wider sense, the reader is referred to the comprehensive exposition in Bachmann [1955] (§§ 40–42) and [1956].

In contrast with the situation at the time of Hausdorff, we no longer expect a proof that inaccessible numbers cannot exist. On the other hand, within suitable systems of axioms it has been shown that, provided the system is consistent, it remains consistent after the addition of an axiom stating that all its cardinal numbers are accessible (cf. Kuratowski [1925], Baer [1929], Firestone–Rosser [1949]). Thus the existence of inaccessible numbers seems to be independent of the other axioms. A special far-reaching axiom, postulating the existence of inaccessible sets (in the narrow sense), was formulated by Tarski [1938] and [1939]. The axiom is strong enough to make the Axioms of Power-Set and of Choice redundant.

As observed above, the formulation of a general postulate of restriction (which would entail the non-existence of inaccessible numbers) seems rather problematic. However, a particular restrictive axiom was introduced by von Neumann [1925] (p. 239) and [1929] (p. 231; cf. Zermelo [1930]) and has proved useful far beyond its original purpose.

Mirimanoff [1917] and [1917/20] was the first to raise the possibility of *ensembles extraordinaires* s_0 with an infinite descending sequence of elements s_k of elements such that

$$\ldots \varepsilon\, s_{k+1}\, \varepsilon\, s_k\, \varepsilon \ldots \varepsilon\, s_2\, \varepsilon\, s_1\, \varepsilon\, s_0.$$

In particular, s_{k+1} may be the only element of s_k, and possibly $s_k = s_l$ for some different k, l, including the extreme case of s_0 containing itself as its only element. This means that primary

---

[1]) The respective ordinals were called *Grenzzahlen* by Zermelo [1930].

constituents (*Urelemente* in Zermelo's terminology, including the null-set) need not exist for every set. The existence of such sets s_0 is consistent with the Axioms **I–VIII** of **Z**, yet it cannot be inferred from them.

Trying to exclude such sets [1]), von Neumann introduced the

> *Axiom* **IX** *of Foundation* [2]). Every non-empty set s contains an element t such that s and t have no common element.

In other words, sets $s \neq O$ such that each element of s has an element which is also contained in s, shall not exist. Hence a descending sequence in the sense mentioned above is always "finite" and "ends" in a primary constituent, which in the case of **Z** is necessarily the null-set. In particular, no set s contains itself as an element; for if $s \, \varepsilon \, s$, $\{s\}$ would contradict Axiom **IX**. However, the efficacy of this axiom still needs further examination.

Finally, the (generalized) *continuum problem* bears upon the question of a maximal limitation, or else a maximal extension, of the domain of set theory. As is well-known, Cantor's efforts of the 1880's, and Hilbert's of the 1920's, to solve the continuum problem have remained unsuccessful. Only in the late 1930's did Gödel ([1938] and [1940]) attain an actual progress by showing that the generalized continuum-hypothesis *cannot be disproved*: it turns out to be consistent with certain axiom systems of set-theory, in particular with Gödel's modification of Bernays' system (cf. below, No. 7). This result, however, raises the question whether the negation of the continuum-hypothesis is consistent as well, which would mean that it were *independent* of the other axioms. Set-theory and analysis would then bifurcate at the continuum-hypothesis in a sense similar to the bifurcation of absolute geometry at the axiom of parallels. Yet, if one conceives the axioms of set-

---

[1]) As pointed out by Gödel [1944], this exclusion bears upon reconciling axiomatic set theory with a simplified theory of types; cf. Quine [1940], §§ 24 and 28.

[2]) *Axiom der Fundierung* (Zermelo [1930]); "restrictive axiom" (Bernays [1941]); also "axiom of grounding".

theory as "describing some well-determined reality", as does Gödel [1947], the hypothesis ought to be either true or false and not independent. Its indecidability on account of known axiom systems might mean that those systems "do not contain a complete description" of that reality, and that there may exist "other (hitherto unknown) axioms of set-theory which a more profound understanding of the concepts underlying logic and mathematics would enable us to recognize as implied by these concepts". Thus the addition of new axioms might be necessary to yield a solution of the continuum problem within the system.

In particular, since the continuum is virtually the set of all subsets of a denumerable set, solving the continuum problem possibly requires a more far-reaching characterization of the concept of subset than obtained above by the Axioms of Subsets and of Choice. What these two axioms furnish, may be separated by a deep abyss from what Cantor had in mind when speaking of subsets of s, viz. arbitrary multitudes of elements of s. One cannot expect to determine the *number* of the subsets of s before it is unambiguously settled *what they are.*

Thus the question of adding further axioms seems to be not only meaningful but even imperative. However, no direction is as yet visible in which such axioms might be sought.

6. Development of set-theory from the axioms of **Z**

The main features of set-theory were deduced, from axioms not much different from those of **Z**, by Zermelo [1908a] (cf. Fraenkel [1925]) as far as the theory of equivalence is concerned, and by Fraenkel [1926] and [1932] for the theory of ordered and well-ordered sets. The concept of order, i.e. the method of reducing the order relation to the membership relation, used in this deduction is that of Kuratowski [1921] (cf. already Hessenberg [1906], ch. 28).

It will be sufficient to give the definitions of equivalence and of (well-) order and to point out a few general and fundamental theorems, which constitute the basis upon which then classical set-theory can be developed along lines which do not essentially differ from the traditional ones.

The equivalence relation can be reduced to the membership relation through the following

Definition of Equivalence. If S and T are mutually exclusive sets, S is called *equivalent* to T ($S \sim T$) if the Cartesian product $S \times T$ (p. 14) has at least one subset ϕ such that every element of $S \cup T$ belongs to a single element of ϕ. Every such ϕ is called a (one-to-one) *mapping* between S and T.

The null-set is called equivalent to itself. Clearly the equivalence relation $\sim$ is symmetrical.

The definition derives from the fact that a mapping between S and T can be regarded as a set of pairs $\{s, t\}$ of elements of S and T such that $\{s_1, t_1\} \neq \{s_2, t_2\}$ implies $s_1 \neq s_2$ and $t_1 \neq t_2$, and that every element of $S \cup T$ appears in (just) one pair.

Naturally, the same way may be taken to reduce the concept of a *single-valued function* $f(s) = t$, with the domain S and the counter-domain T, to the membership relation. Since then the correlation need not be one-to-one, the last condition of the above definition will be weakened correspondingly.

The set of all mappings between mutually exclusive sets A and B exists within **Z**. For according to Theorem 3 (p. 14) and Axiom **IV** the power-set $\Pi(A \times B)$ exists, and mappings between A and B, if any, are certain elements of this power-set. Therefore, taking the condition on ϕ expressed in the above definition as the predicate appearing in the Axiom of Subsets (p. 11), this axiom yields a set each of whose elements ϕ is a mapping as desired. Hence $A \sim B$ if and only if this set is not empty.

The concept "image in T of a given $s \varepsilon S$ with respect to a given mapping" is easily obtained. We get rid of the restriction to mutually exclusive sets by means of the following theorem [1]), which is proven without the Axiom of Choice: *To every given set D there corresponds a disjointed set D' such that $D' \sim D$ and that, d' denoting the image in D' of $d \varepsilon D$ with respect to a suitably chosen mapping between D and*

[1]) This theorem is much stronger than required for the present purpose, but necessary for proving the general principle of choice (see p. 28).

D', one has $d' \sim d$ for every $d \varepsilon D$. Moreover, D' can be chosen such that $\bigcup D$ and $\bigcup D'$ are mutually exclusive. Accordingly, the equivalence between sets S and T which are not mutually exclusive can be defined by means of a set U which has no common element with S or T and is equivalent to both sets.

It follows that Zermelo's general principle of choice (p. 16) can be deduced from Axiom **VI** (p. 15). For starting from any set S of non-empty sets, we can form an equivalent *disjointed* set s whose elements are respectively equivalent to the elements of S in the sense just specified. A selection-set σ of s exists on account of Axiom **VI**, and from the elements of σ we pass on to corresponding elements of the elements of S by simultaneous mappings between the elements of S and those of s. Thus to each element of S one of its elements is assigned.

Cantor's, Schröder–Bernstein's, and König–Zermelo's theorems are now proven in the usual way, the latter by means of the Axiom of Choice. The necessity of avoiding cardinal numbers presents no actual difficulty though it makes certain proofs somewhat tedious.

The order relation may be reduced to the membership relation by the following

Definition of Order (ordered set)[1]. A set S is called orderable if a maximal chain M of subsets of S exists, i.e. if the power-set $\Pi(S)$ has at least one subset M with the following properties

a) if m and m' are different elements of M, either $m \subset m'$ or $m' \subset m$;

b) if $\overline{M}$ is a subset of $\Pi(S)$ with the property a) and if $M \subseteq \overline{M}$, then $M = \overline{M}$.

M is called *an order of* S.

This definition is based on considering M to be the set of *all initials* (or all remainders) of an ordered set S. In fact, the set I of all initials has the "chain"-property that, of two different initials,

[1]) We are only dealing with "total" or "complete" order, not with partially ordered sets.

one is a proper initial—*a fortiori*, a proper subset—of the other. Moreover, given two elements of S, there is at least one initial that contains *just one* of these elements, and for every subset I_0 of I, also $\bigcup I_0$ and $\bigcap I_0$ are initials. Finally, O and S are initials of S.

These properties together prove *characteristic* of I (or of the set of all remainders), among all sets of subsets of S. Kuratowski [1921] now proved the remarkable fact that, besides the chain-condition, the conjunction of these properties is equipollent to the property b) which demands I to be a *maximal* chain.

An order M in the above sense turns out to be an order of S in the traditional sense, if $s_1 \prec s_2$ is understood to mean that there exists an element of M (i.e., an initial) which contains s_1 but not s_2. The chain-condition involves the asymmetry and the transitivity of $\prec$.

To decide whether a given set S can be ordered, we form the set of *all* orders M of S, which exists on account of Axioms **I**–**V**. To prove this, we start from the power-set $\Pi(\Pi(S)) = K$, whose elements x are *sets of subsets* of S. Then those x which satisfy the condition of being a maximal chain form a subset K_0 of K on account of the Axiom of Subsets, and the elements of K_0 are all possible orders of S. Hence S is orderable or not, according as K_0 is $\neq O$ or $= O$.

By means of the Axiom of Choice, to be sure, it can be shown that every set is orderable; for instance, through the well-ordering theorem, whose second proof in Zermelo [1908] can easily be imbedded in **Z**, or through the theorem of Szpilrajn [1930] according to which every partial order of a set can be extended to a complete order.

Another method of reducing the order relation to the membership relation is that of Hausdorff [1914], pp. 70f. Here an order of S is considered to be the set of all those ordered pairs (s_1, s_2) with $s_1 \varepsilon S$, $s_2 \varepsilon S$, for which $s_1 \prec s_2$. The ordered pair (a, b) is, as usual, defined as the (unordered) pair $\{\{a, b\}, \{a\}\}$.

Starting from the definition of order, we may develop the theory of ordered sets along traditional lines. As in the theory of equivalence, the set of all similar mappings between two ordered sets can be shown to exist on account of Axioms **I–V**; if this set is not empty, the sets are called *similar.*

To establish the theory of *well-ordered* sets one need not go back to the general concept of order. On account of the Definition of Order, for any subset M_0 of an order M of S we have $\bigcap M_0 \varepsilon M$ (cf. p. 29). Sharpening this to the condition $\bigcap M_0 \varepsilon M_0$, we obtain the additional characteristic property of well-ordering; for $\bigcap M_0 \varepsilon M_0$ means that M_0 has a minimal element.

By also using the Axiom of Substitution (p. 22) we can develop in **Z** von Neumann's [1923] theory of ordinal numbers, where the ordinals appear as special sets defined by transfinite induction (cf. von Neumann [1928a]). Then the ordinals o prove to be the well-ordered sets with the property that every $x \varepsilon o$ is also the section (segment, *Abschnitt*) of o determined by x. It follows that, of two different ordinals, one is both an element and a proper subset of the other, which property yields the natural order of ordinals "according to their magnitude". It can be shown that to every well-ordered set w corresponds just one ordinal similar to w; "its" ordinal. Therefore also the *cardinal numbers* (Alefs) can be introduced explicitly, namely as the initial numbers of the various number-classes; in other words, a cardinal is an ordinal c no element of which is equivalent to c.

The existence of an order to any given set was established above through the Axiom of Choice or the (equipollent) well-ordering theorem. This, however, is more than desired, inasmuch as well-ordering means imposing an order of a special structure. Hence if, instead of Axiom **VI** on p. 15, we postulate the ordering theorem, viz. that every set can be ordered, it seems that we obtain a weaker system. One easily proves that the ordering theorem cannot be proved by Axioms **I–V** (at least if infinitely many individuals may be admitted to the system); for the theorem implies (Fraenkel [1928a]) the weakest axiom of choice, which itself is independent (cf. above pp. 18/9). However, whether this implication holds also in the converse sense is as yet unknown. On the other hand, Mostowski [1939] (cf. Doss [1945]) proved that the ordering theorem is indeed weaker than the well-ordering theorem (above p. 19). Hence, by adding the ordering theorem to Axioms **I–V**, **VII**, **VIII**, we obtain a system less strong than **Z**.

The theories of finite and of denumerable sets are easily obtained by specialization. As to finite sets, which for instance may be considered to be "doubly well-ordered sets", one can dispense with the Axiom of Infinity, while the Axiom of Choice is required for certain problems (p. 19). Denumerable sets are the sets equivalent to the set W_0 of p. 21; their fundamental properties are derived from the axioms of **Z** in Zermelo [1908a].

While **Z** thus turns out to be sufficient for developing classical set theory, the antinomies known hitherto cannot appear in **Z**, as hinted on p. 8 and p. 11.

7. Remarks on the Axiom Systems of von Neumann, Bernays, Gödel

The double purpose of establishing classical set-theory and of avoiding the antinomies which show Cantor's attitude to be impracticable, has been answered by the system **Z**, and to a great extent already by Zermelo's [1908a] original set of axioms. The rather arbitrary character of the processes which are chosen in the axioms of **Z** as the basis of the theory, is justified by the historical development of set-theory rather than by logical arguments. The far-reaching aim of proving the consistency of **Z**, which would exclude contradictions of types as yet unknown, is not likely to be attained at the present stage, and in a well-defined sense cannot be attained at all, in accordance with Gödel's incompletability theorem.

While therefore satisfactory in many respects, **Z** has nevertheless an essential disadvantage. Certain sets, among them virtually all that are required for the classical theory, can be proved to exist, and others (e.g., those which yield the well-known antinomies) are easily proved not to exist within **Z**. However, between these domains of existence and of non-existence a vast field of uncertainty still remains; and what is more important, the bulk of this field consists of concepts—notably, very comprehensive sets such as the set of all sets, the set of all ordinals, etc—which are apt to lead to contradiction not through their being conceived as sets (classes) but through operations with them, in particular through their being taken as *elements* of sets. For themselves these concepts prove quite harmless. Thus it turns out that the demarcation line drawn in **Z** between the sets whose existence can be proved and other sets is rather arbitrary.

A further weakness of **Z** is that the Axiom of Subsets (No. 3) constitutes not a proper axiom but an axiom schema (p. 12/13), and the same applies to the Axiom of Substitution (No. 5). It would be preferable to get rid of axiom schemata, and hereby to render the system "elementary" in the sense that all axioms can be formulated within the predicate calculus of first order whose bound variables are restricted to individuals (in the logical sense, excluding predicates).

All these tasks are accomplished in *von Neumann's foundation of set theory* ([1925, 1928, 1929], simplified in R. M. Robinson [1937]). Very considerable improvements and additions to this foundation, together with a fundamental simplification, are due to Bernays [1937–54]. Finally, a modification of Bernays' system is given in Gödel [1940], chiefly for the purpose of proving the consistency of the generalized continuum-hypothesis.

Exhibiting the main features of these systems would require a lengthy exposition. This is superfluous at this juncture since the main part of the present monograph is just a modified and improved elaboration of ideas which are found in Bernays' papers mentioned above. Therefore the only purpose of the present section is to point out the principles and tendencies of von Neumann, Bernays, Gödel *in comparison with the system* **Z** exhibited thus far. We shall, however, ignore von Neumann's peculiar terminology.

Since "over-comprehensive sets" do only harm by being taken as elements of other sets, a distinction is made between collections that can also serve as elements, henceforth called *sets*, and collections excluded from elementhood, called *classes*. While to every set corresponds a definite class, namely that which contains the same elements [1]), the converse does not hold; in general no set corresponds to a given class. Strictly we have therefore to distinguish between two primitive membership relations ε and η; $x \varepsilon s$ applies to a set s, $x \eta C$ to a class C. (Gödel dropped the distinction between ε and η for the sake of simplicity.) Classes may be "over-comprehensive"; there exists, for instance, the universal class $\vee$, defined by the identical predicate $x = x$, and the complement $\vee - s$ of a given set (or class) s. Therefore the main limitation contained in the Axiom of Subsets (p. 11), viz. the selection of elements from *a set already secured* according to a given predicate, is dropped; every well-formed formula (predicate) defines a class, which contains all objects (sets) that satisfy the predicate. Moreover, the distinction between sets and classes can be expressed exactly by the over-comprehensiveness of the latter. In fact, von Neumann introduced an axiom – his IV2, most characteristic of his system – which roughly excludes from elementhood (i.e., admits as classes but not as sets) just those collections which can be mapped on the universe $\vee$; for instance, the class of all ordinal (or cardinal) numbers.

However, while for von Neumann this axiom also includes the Axioms of Subsets, of Choice, and of Substitution, Bernays took an essential step forward by distinguishing between these different purposes. The Axiom of Choice is retained without an essential change. Yet with respect to the Axiom of Subsets, an ingenious device enables Bernays to dispense with the general concept of predicate or function in his *class axiom*, which correspondingly to Axiom **V** of **Z** would read: For every predicate F which contains no bound class-variables, there exists the class of those sets which satisfy F. This seems to be an axiom schema, but is (in Bernays' axiom group III) reduced to a small number of single axioms for the construction of classes. Hereby also the Axiom of Substitution can be transformed to a single axiom. (In von Neumann's system the same purpose is attained in a different way.)

---

[1]) In Gödel's system, the class is even identified with the set to which it corresponds.

A secondary, but practically important feature is that the ordinal numbers are explicitly introduced at an early stage.

As to Gödel's system, its significance consists less in various innovations and simplifications of the formal system than in the construction, within his axiomatic system, of a certain *model* Δ, which shows the consistency of the Axiom of Choice and of the generalized continuum-hypothesis. The sets of Δ are the elements of the class L of values of a certain function whose domain is the class of all ordinals; L satisfies a *postulate of constructibility* which is not incorporated in the axiom system itself, though it proves compatible with the axioms (cf. Gödel [1947]). The term "constructible" is here taken in a loose sense, roughly meaning that, starting with the null-set, we may iterate the application of the operations of the system any transfinite number of times. In particular, every ordinal proves constructible.

While the domain of sets existing in view of **Z** is limited in a somewhat arbitrary way, the domain of von Neumann–Bernays–Gödel – which comprehends the sets ensured in **Z** as a small portion – is as extensive as seems possible in view of the antinomies, if inaccessible numbers (sets) are disregarded. Therefore it is remarkable that, *on the assumption that* **Z** *is consistent, also the system of Bernays–Gödel can be proven to be consistent* [1]). In an essential point these relative consistency proofs rest upon the fact that classes are defined in a predicative way, as against the impredicative procedure of Axiom **V** of p. 11; the proof of Rosser and Wang uses the well-known theorem of Löwenheim and Skolem which guarantees the existence of a model in the domain of positive integers. According to Mostowski, in particular every theorem of Gödel's system which involves *set*-variables only, is also a theorem of **Z** [2]).

Finally it should be stressed that, if $\overline{\mathbf{Z}}$ denotes the system of our Axioms **I–VII**, Axiom **VIII** is independent of $\overline{\mathbf{Z}}$, and $\overline{\mathbf{Z}}$ is a weaker

---

[1]) Shown independently and by different methods in Novak [1948/51] (cf. Mostowski [1951] and McNaughton [1953]) and Rosser–Wang [1950]; improved in Shoenfield [1954].

[2]) Cf. Rosser–Wang [1950], pp. 125–128.

system in a strong sense than that of Axioms **I–VIII** [1]). (While by **I–VII** only the existence of denumerably many transfinite cardinals is ensured, by means of **VIII** all up to the first inaccessible number are added. Informally one may say that through the classes only a single "cardinal", viz. that of the universal class, enters into the system.) One may conclude that the system of Bernays and **Z** prove to have somehow comparable strength, as in contrast with $\overline{\mathbf{Z}}$ and **Z**.

---

[1]) Cf. Wang–McNaughton [1953], p. 18; Montague [1956].

PART II

AXIOMATIC SET THEORY

BY

P. BERNAYS

INTRODUCTION

Axiomatic set theory has first been set up by Ernst Zermelo in order to deal with the set-theoretic antinomies. The discussion of these antinomies has still not lost its actuality; even there is in the discussions often the tendency to exaggerate the consequences of them. Of course the importance of these antinomies is beyond any doubt. But what may be regarded as an exaggeration, is to infer from them a requirement of restricting our usual methods of mathematical procedure. In particular no cogent argument can be drawn from them that mathematics have to be built up in a strictly predicative way,–if one is not a priori convinced of the necessity of a predicative procedure.

What really is excluded by the antinomies is only that interpretation (easily suggested at first) of set theory which finds its formal expression in the calculus called newly Idealkalkül by Hermes and Scholz [1952], whose domain of individuals contains for every predicate $\mathfrak{P}$ an assigned individual p such that

$$(x)(x \in p \leftrightarrow \mathfrak{P}(x)).$$

The impossibility of this interpretation shows that the interrelation between logic and mathematics cannot be characterized in a too simple way, but a nearer delimitation is required. This delimitation however can be given in such a way that our usual mathematical procedures of formation and inference are not restricted, and to do this by the axiomatic method in a way as suitable as possible is the task of axiomatic set theory.

For a similar purpose the systems of mathematical logic have been edified, first the famous Principia Mathematica [Whitehead and Russell 1925/27] and in our times the systems of W. V. Quine [1937, 1951]. There is a development from the one to the others, leading over the simple theory of types.

In the P. M. there is a manifold of tendencies which partly conflict with one another. In fact the formalism is first constructed according to a device of predicativity which however in the following is not maintained, so that on the whole the same result can also be attained by the easier simple theory of types, with the axiom of infinity included. Still the distinction of types here present amounts to a complication which is somewhat cumbersome for the mathematician.

In particular W. V. Quine has pursued the design of removing the type distinction. In his system of mathematical logic, to which B. Rosser and H. Wang contributed, only a residual of this distinction, the condition of "stratification", figures as a directive device for regulating the connection between predicate formation and set formation. By this way the arrangement comes very near to that of axiomatic set theory. This appears the more if in set theory the extensions of the predicates are introduced as a second kind of individuals, as first von Neumann has done. In fact also in Quine's system those individuals which are elements are distinguished from the others.

The use of the criterium of stratification has a clearly experimental character. There are also some strange features of the system induced by it. Thus here the existence of sets or classes having themselves as elements is not only not excluded (as in other forms of set theory) but even provable. Further it can here be inferred that the class of all individuals is not in a one-to-one correspondence with the class of unit sets. Other anomalies have appeared by a recent publication of E. Specker [1953].

The system here presented follows nearer the line of Zermelo's axiomatics. From the development of axiomatic set theory in the meantime it adopts some devices. The modifications thus arising can be characterized by the following steps:

1) The Zermelo concept of "definite Eigenschaft" is precised by the Skolem method, i.e. using the formal language of logic. By this way it becomes possible to formalize the system in the frame of the predicate calculus of first order with admitting besides formal axioms also axiom schemata.

2) We adopt the generating process afforded by the Fraenkel Ersetzungsaxiom. This axiom was originally stated by Fraenkel

[1922] with a certain reserve, but it is a natural supplement to Zermelo's axiomatics. In fact, in combination with the Ersetzungsaxiom Zermelo's sum axiom yields Cantor's summation process in its full generality. Indeed for the Cantor summation it is not essential that the sets to be summed are the elements of a set, but only that the summation index ranges over a set. As to the concept of function which occurs in the premise of the Ersetzungsaxiom, its precising is already supplied by that one of definite Eigenschaft, in connection with the definition of an ordered pair, which we adopt in the form of Kuratowski [1921].

3) From von Neumann's axiomatics [1925, 1928] we adopt first the idea of embodying in the system a part of formalization of the metamathematics – however keeping nearer to the former systems by retaining the original binary element relation instead of von Neumann's ternary relation "the function f has for the argument x the value y"; secondly we follow von Neumann in separating the rôle of the potency axiom, as one of the stronger axioms, from the conceptual set formation, which can be made with the aid of those axioms which are directly related to the concept of definite Eigenschaft.

In the axiomatization given in the Journ. of Symb. Log. [Bernays 1937–1954] still the further von Neumann device was adopted of setting up the system as a Sortenkalkul of first order with two kinds of individuals called here sets and classes. By this method it becomes possible to avoid the use of axiom schemata so that a system with only finitely many formal axioms results. The two kinds of individuals, as well known, can on principle be reduced to only one kind, so that we come back to a one-sorted system. Here in particular this can be done in a special simple way, namely by taking as sets those classes which satisfy the condition to be an element of a class. This method was applied by Tarski, Mostowski and Gödel; it occurs also in Quine's mathematical logic.

However it might be asked if we have here really to go so far in the formal analogy with usual axiomatics. Let us regard the question with respect to the connection between set theory and extensional logic. As well known, it was the idea of Frege to identify sets with extensions (Wertverläufe) of predicates and to treat these extensions on the same level as individuals. That this idea cannot be maintained was shown by the Russell paradox.

Now one way to escape the difficulty is to distinguish different kinds of individuals and thus to abandon Freges second assumption; that is the method of type theory. But another way is to give up Frege's first assumption, that is to distinguish classes as extensions from sets as individuals [1]). Then we have the advantage that the operation of forming a class $\{x \,\|\, \mathfrak{A}(x)\}$ from a predicate $\mathfrak{A}(c)$ can be taken as an unrestricted logical operation, not depending on a specifying comprehension axiom. But from this then we have to separate the mathematical processes of set formation which in the way of Cantor are performed as generalizations of our intuitive operations on finite collections.

The task of axiomatic set theory then consists in showing that the method of this generalization can be precised in such a way that on one hand we have the full freedom of set-theoretic formations and a sufficient frame for classical mathematics and on the other hand do not fall into the antinomies. From this point of view we shall not tend to bring formally together classes and sets in one domain of individuals and even we might refrain at all from handling with a domain of classes to which bound variables refer. We might remind here of the criticism which Quine [1941a] applied to the use of bound variables relating to predicates in P. M., pointing to the distinction between propositional functions and attributes. We want to have the classes related to propositional functions, – though not in a way of contextual definition but by a general rule of conversion which we regard as belonging to the logical grammar. Thus we are induced to employ class variables only as free variables, what in fact is sufficient for giving our

---

1) An other proposal for overcoming the difficulty of the antinomies was made by Heinrich Behmann [1931], which consists in admitting the possibility that the logical functions for some arguments yield a senseless value. Here senseless expressions are formally allowed, what makes modifications of elementary logic being required. The program of a formal system according to this device is till now not yet accomplished.

formal language a suitable flexibility, so that we are not obliged to employ continually syntactical variables in the formulations.

At the same time this formally stronger separation of the classes from sets conforms to the view that the universe of mathematical objects (sets) is not itself a mathematical object A further circumstance by which the restriction of class variables to free variables is suggested, is that in [Bernays 1937–1954] each of the axiomatic statements of class existence – with exception of that in the axiom of choice – is equivalent to an explicit introduction of a class formation, and the axiom of choice, as there stated, can for the usual mathematical purposes be replaced by an assertion of mere set existence.

Besides it might be noted that in [Bernays 1937–1954] the bound class variables never explicitly occur in symbolic notations but only verbally in statements by language. In the present monograph the formalization is carried out to a higher degree. So set-theory is here presented as a formal system, and also a basic logical frame is delimited so that formal derivations can be described. This stronger formalization is for the purpose to conform to the want to-day often existing of having formalizations at hand in axiomatic theories and at the same time to facilitate the comparison with the comprehensive systems of mathematical logic.

There are some further differences against [Bernays 1937–1954]. In stating the axioms we avoid the existential form by using primitive symbols. (The axiom of choice will be the only axiom stated in existential form.) The succession of steps in the edification of the theory is somewhat different. In particular we here begin the proper set-theoretic development with three axioms parallel to the generating processes of Cantor's ordinal arithmetic; these already are sufficient for obtaining the general ordinal theory, including the full recursion procedures.

In this monograph we shall give the axiomatic edification with discussions of the axioms and with their applications yielding classical mathematics, especially ordinal theory, number theory, analysis and cardinal theory. The further axiomatical questions on eliminability, relative consistency and independence are left to a second volume.

Remark on the indication of formulas and definitions: The monograph is divided in chapters and sections. To these the numeration of formulas and definitions refers in the way that signed formulas and definitions have three numbers, the first (a roman number) referring to the chapter, the second (an arabic number) to the section in that chapter and the third (also arabic) to the succession within this section. In quotations referring to the same chapter the roman number will be omitted.

CHAPTER I

THE FRAME OF LOGIC AND CLASS THEORY

I, § 1. Predicate calculus; class terms and descriptions, explicit definitions

Our logical frame is an extension of that elementary logical calculus which is called the Prädikatenkalkul or quantification theory. This extension consists in adding descriptions and besides a formalism of classes. Both these extensions are on principle eliminable, as we shall show in the second volume.

Let us now enumerate our rules of formation and derivation. We have free and bound set variables and free class variables. For the set variables we take small latin letters, for the class variables capital italics. For bound set variables we reserve the letters x, y, z, u, v, w, in distinction from free set variables. From the variables belonging to the formal system we have to distinguish the syntactical variables denoting formulas or parts of formulas. As syntactical variables we employ small and capital german letters; small german letters stand for set terms, the capital german letters $\mathfrak{A}$, $\mathfrak{B}$, $\mathfrak{C}$, $\mathfrak{D}$ for formulas and $\mathfrak{F}$, $\mathfrak{G}$, $\mathfrak{H}$, $\mathfrak{K}$, $\mathfrak{L}$, $\mathfrak{M}$, $\mathfrak{N}$ for class terms. In all cases the letters can have with them one or more arguments. The letters $\mathfrak{x}$, $\mathfrak{y}$, $\mathfrak{z}$, $\mathfrak{u}$, $\mathfrak{v}$, $\mathfrak{w}$ will be reserved for the syntactical notation of bound set variables.

Our logical symbols are:

1) The propositional connectives

conjunction	&	«and» [1]
negation	$-$	«not»
alternative	$\vee$	«or», in the sense of "or else"
implication	$\rightarrow$	«if ... so»
biimplication [2]	$\leftrightarrow$	«if and only if»

---

[1] Formulations between "«" and "»" are meant to explain the interpretation of symbols or formulas, in a more or less free way.

[2] We take this word in order to avoid the terminus "equivalence" which indeed is used mostly in other senses.

2) The quantifiers
the general quantifier $(\mathfrak{x})\mathfrak{A}(\mathfrak{x})$ «for all $\mathfrak{x}$, $\mathfrak{A}(\mathfrak{x})$»,
the existential quantifier $(E\mathfrak{x})\mathfrak{A}(\mathfrak{x})$ «for some $\mathfrak{x}$, $\mathfrak{A}(\mathfrak{x})$»,
3) The class operator $\{\mathfrak{x} \mid \mathfrak{A}(\mathfrak{x})\}$ [1]) «the class of the $\mathfrak{x}$ such that $\mathfrak{A}(\mathfrak{x})$»,
4) The description operator (ι-symbol) $\iota_{\mathfrak{x}}(\mathfrak{A}(\mathfrak{x}), a)$ «the set $\mathfrak{x}$ such that $\mathfrak{A}(\mathfrak{x})$, or else a».

In all these cases $\mathfrak{x}$ is a bound variable and $\mathfrak{A}(\mathfrak{x})$ is the expression resulting from a formula $\mathfrak{A}(\mathfrak{a})$ with a free variable $\mathfrak{a}$, wherein $\mathfrak{x}$ does not occur, by replacing everywhere $\mathfrak{a}$ by $\mathfrak{x}$.

By means of the propositional connectives and the quantifiers formulas are built up in the familiar way out of prime formulas. A prime formula consists of a predicate symbol (predicator) with terms as arguments. Among the terms we distinguish set terms and class terms; of both kinds of terms we have the following possible cases:

1) a free set- or class variable,
2) an individual symbol (set- or class- symbol) [2]),
3) a function symbol (functor) with terms as arguments,
4) an expression formed by the application of an operation symbol (operator) to an arbitrary formula or term, of which one or more occurring free variables are bound by the operation symbol.

By the occurrence of formulas as constituents of terms the definition of the concepts of term and formula become somewhat involved, in the way that they have to be defined in common, what however can precisely be done in a recursive manner.

In order to indicate the scopes of connectives, quantifiers, functors and operators within a formula, we are employing brackets

---

[1]) This symbol (used in some newer papers) is here taken, for the sake of easier print, instead of Russell's class symbol $\hat{\mathfrak{x}}\mathfrak{A}(\mathfrak{x})$, whose adoption was first intended.

[2]) The word "individual symbol"—used for set- or class symbols—is here taken for denoting a syntactical category. It is not meant by this that classes in the interpretation have to be regarded as individuals. The way of interpreting the class formalism will be nearer discussed in I, § 3.

and parentheses as usual in mathematics, and also dots for sparing parentheses. Further for avoiding parentheses or dots we agree that the symbols $\rightarrow$ and $\leftrightarrow$ have the preference before & and $\vee$ as to the separation of expressions.

Now we come to the *rules of derivation*. The formal derivative development of a theory goes by formal inferences starting from initial formulas. It could seem that the rôle of these two parts of the derivation process should be so that by the initial formulas the axioms of the theory and by the rules of inference the logical reasonings have to be formalized. However the separation of the two parts is not so strict since in the formal arrangement it is often suitable to express logical rules by means of formula schemata.

It is not essential for our purpose to distinguish one way of performing the formalizing of logical reasonings. In fact there are various elegant ways of shaping the predicate calculus. Only for fixing our procedure we adopt here a definite form of this calculus, which is mainly that of [Hilbert and Bernays 1934/39].

We take as initial formulas:

1) The propositional identities, i.e. formulas composed out of partial formulas by means of the propositional connectives in such a way that for any assignment of arbitrary truth values to the partial formulas we obtain by the truth table method the value "true".

2) Formulas conforming to one of the schemata

$$(\mathfrak{x})\mathfrak{A}(\mathfrak{x}) \rightarrow \mathfrak{A}(\mathfrak{t}), \qquad \mathfrak{A}(\mathfrak{t}) \rightarrow (E\mathfrak{x})\mathfrak{A}(\mathfrak{x}),$$

where $\mathfrak{t}$ is a term and $\mathfrak{A}(\mathfrak{t})$ results from $\mathfrak{A}(\mathfrak{x})$ by replacing everywhere $\mathfrak{x}$ by $\mathfrak{t}$.

As rules of inference we take:

1) The modus ponens

$$\frac{\mathfrak{A},\ \mathfrak{A} \rightarrow \mathfrak{B}}{\mathfrak{B}},$$

2) The schemata for the quantifiers

$$\frac{\mathfrak{A} \rightarrow \mathfrak{B}(a)}{\mathfrak{A} \rightarrow (\mathfrak{x})\mathfrak{B}(\mathfrak{x})}, \qquad \frac{\mathfrak{B}(a) \rightarrow \mathfrak{A}}{(E\mathfrak{x})\mathfrak{B}(\mathfrak{x}) \rightarrow \mathfrak{A}},$$

with the condition that the free variable $\mathfrak{a}$ does not occur in $\mathfrak{A}$ and neither in $\mathfrak{B}(\mathfrak{x})$, and that $\mathfrak{x}$ does not occur in $\mathfrak{B}(\mathfrak{a})$.

The application of syntactical letters for bound variables is to the effect to make dispensable a special rule of Umbenennung for the bound variables as fundamental rule. The same remark refers also to the like application of syntactical letters in further schemata and definitions of operators.

From these rules in connection with the initial formulas all the schemata of inferences of the predicate calculus are obtainable as derivable schemata. We note in particular:

$$\frac{\mathfrak{A} \to \mathfrak{B},\ \mathfrak{B} \to \mathfrak{C}}{\mathfrak{A} \to \mathfrak{C}},$$

«Kettenschluss»

$\mathfrak{A}\ \&\ \mathfrak{B} \to \mathfrak{C}$ is replaceable by $\mathfrak{A} \to (\mathfrak{B} \to \mathfrak{C})$ and inversely,

«exportation and importation»

$$\frac{\mathfrak{A}(a)}{\mathfrak{A}(\mathfrak{t})},$$

«rule of substitution»

$$\frac{\mathfrak{A}(a)}{(\mathfrak{x})\mathfrak{A}(\mathfrak{x})}.$$

We are now to extend the predicate calculus by the rules on the class- and ι-operators. In these rules the predicators "=" and "$\in$" are occurring which are the only primitive predicators of our system. The primitive prime formulas formed with them are exclusively the following:

1) $\mathfrak{a}=\mathfrak{b}$, «$\mathfrak{a}$ equal $\mathfrak{b}$», with $\mathfrak{a}$ and $\mathfrak{b}$ being set terms,
2) $\mathfrak{a} \in \mathfrak{b}$, «$\mathfrak{a}$ element of $\mathfrak{b}$» or «$\mathfrak{a}$ in $\mathfrak{b}$», with $\mathfrak{a}$ and $\mathfrak{b}$ being set terms,
3) $\mathfrak{a} \in \mathfrak{K}$, «$\mathfrak{a}$ element of $\mathfrak{K}$» or «$\mathfrak{a}$ belongs to $\mathfrak{K}$», with $\mathfrak{a}$ being a set term and $\mathfrak{K}$ a class term.

In our handling with equality and the element relation the assumption is implicitly formalized that for any sets $\mathfrak{a}$, $\mathfrak{b}$ and for any set $\mathfrak{a}$ and any class $\mathfrak{K}$ it is uniquely determined if $\mathfrak{a} = \mathfrak{b}$ or not and if $\mathfrak{a}$ is an element of $\mathfrak{b}$, resp. of $\mathfrak{K}$, or not. – We write $\mathfrak{a} \neq \mathfrak{b}$, $\mathfrak{a} \notin \mathfrak{b}$, $\mathfrak{a} \notin \mathfrak{K}$ for the negations of the said prime formulas.

The rules concerning the class terms are: the formula schema ("Church schema")

$$c \in \{\mathfrak{x} \mid \mathfrak{A}(\mathfrak{x})\} \leftrightarrow \mathfrak{A}(c),$$

expressing a conversion law, in the sense of A. Church [1932], and the substitution rule (parallel to that one for set variables)

$$\frac{\mathfrak{A}(C)}{\mathfrak{A}(\mathfrak{K})}$$

where $\mathfrak{K}$ is a class term. Of course it is meant that here C is replaced by $\mathfrak{K}$ at all places of its occurrence in $\mathfrak{A}$.

With regard to the formal handling with descriptions there are the well known difficulties, which recently again have been discussed by Carnap [1947]. Following usual language one would be inclined to admit an expression $\iota_{\mathfrak{x}}\mathfrak{A}(\mathfrak{x})$ as a term only if the unicity formulas $(E\mathfrak{x})\mathfrak{A}(\mathfrak{x})$ and $(\mathfrak{x})(\mathfrak{y})(\mathfrak{A}(\mathfrak{x})\ \&\ \mathfrak{A}(\mathfrak{y}) \rightarrow \mathfrak{x}=\mathfrak{y})$ are proved, as it has been done in [Hilbert and Bernays 1934/39]. To this measure there has been objected that the concept of term is complicated by it in a way that in general there is no decision procedure for what expressions are terms. There is even something further to be objected to this restrictive measure—not yet mentioned, as it seems, in the literature—namely that it gives a difficulty in the case that the unicity formulas are derivable only upon a condition, to be introduced as an antecedent. On the other hand, if for every formula $\mathfrak{A}(c)$ the expression $\iota_{\mathfrak{x}}\mathfrak{A}(\mathfrak{x})$ is admitted as a term, then it is desirable to have the ι-terms formally determined by the rules and axioms in a way that no trivially undecidable formulas arise. The normation required for this determination is performable using an extralogical individual symbol. The extralogical reference can be avoided by introducing a parameter in the ι-term, which in the application to any special theory can be replaced by a definite individual symbol. Conforming to this device we come to the following two formula schemata for the ι-operator, yielding a set term $\iota_{\mathfrak{x}}(\mathfrak{A}(\mathfrak{x}), a)$ from any formula $\mathfrak{A}(c)$:

$$\mathfrak{A}(c)\ \&\ (\mathfrak{x})\,(\mathfrak{A}(\mathfrak{x}) \rightarrow \mathfrak{x}=c) \quad \rightarrow \quad c=\iota_{\mathfrak{x}}(\mathfrak{A}(\mathfrak{x}), a),$$

$$(E\mathfrak{x})\,(\mathfrak{A}(\mathfrak{x})\ \&\ (\mathfrak{y})\,(\mathfrak{A}(\mathfrak{y}) \rightarrow \mathfrak{x}=\mathfrak{y})) \quad \cdot\mathbf{v}\cdot \quad \iota_{\mathfrak{x}}(\mathfrak{A}(\mathfrak{x}), a)=a.$$

By the schemata for class terms and descriptions our repertory of initial formulas is enlarged. A further enrichment is effected by the rule of setting up explicit definitions.

By explicit definitions predicators, functors or operators can be introduced. The definition is given by an initial formula or formula schema, consisting of a left side expression, a right side expression and a connecting symbol. The left and the right side are always either both formulas or both set terms or both class terms. The free variables are the same on the left and on the right. The left side consists only of a symbol with its arguments, of which however some can be syntactical variables having again arguments, which then are bound variables on which the symbol operates and which are marked as indices of the symbol.

The connecting symbol could be a common separate symbol for definitory equality. We then should have a general rule of replacing the definiendum by the defining expression. In order that this rule be not too complicate we should have to make use of Church's λ-notation.

Here we proceed in the way to make the replacement going automatically by rules already adopted: we take as connecting symbol the biimplication $\leftrightarrow$, or the equality symbol $=$, or a predicate symbol, itself still to be defined, of class equality $\equiv$, according as the both sides of the definition are formulas or set terms or class terms. This procedure has the advantage that by the connecting sign in a definition the character of the defined symbol is made directly to appear.

There is still one characteristic of an explicite definition, which includes a reference to a definite order of the definitions in a system, namely that the symbols occurring on the right side are either primitive symbols of the system or already introduced by former definitions. This condition in particular implies that the symbol on the left side does not occur on the right side [1]).

We shall mark definitions by the sign Df connected with the number of the definition. – Let us state at once the definition

[1]) There would be the possibility of making the order of definitions explicit by using as defined symbols exclusively numbered symbols, but the advantage of this procedure is not so high for the direct handling in the formal system as for its metamathematical investigation. However for this purpose we have at all events the method of Gödel numbering at our disposal.

of class equality

$$A \equiv B \leftrightarrow (x)\,(x \in A \leftrightarrow x \in B).$$

Formally an explicit definition in our system has the character of an axiom or axiom schema but with the property that by its special form an easy way of eliminating the defined symbol is available, so that no proper assumption is formalized by it.

For the embodying of particular mathematical theories, as the theory of real numbers or of ordinal numbers, in our set theory, it will be advisable to apply the method of using specialized variables. Such a new kind of free and bound variables—let us denote them by $\mathfrak{a}_1, \mathfrak{b}_1, \mathfrak{c}_1 \ldots \mathfrak{x}_1, \mathfrak{y}_1, \mathfrak{z}_1$ respectively—is associated with a class $\{\mathfrak{x} \,|\, \mathfrak{D}(\mathfrak{x})\}$ over which the variables are ranging.

The rules for introducing these variables are:

1) Every new free variable is a term.
2) If $\mathfrak{C}(\mathfrak{a})$ is a formula with the free variable $\mathfrak{a}$, wherein the bound variable $\mathfrak{x}_1$ does not occur, then $(\mathfrak{x}_1)\mathfrak{C}(\mathfrak{x}_1)$, as also $(E\mathfrak{x}_1)\mathfrak{C}(\mathfrak{x}_1)$ is a formula and $\{\mathfrak{x}_1 \,|\, \mathfrak{C}(\mathfrak{x}_1)\}$ is a class term and $\iota_{\mathfrak{x}_1}(\mathfrak{C}(\mathfrak{x}_1), a)$ is a set term.
3) From a formula $\mathfrak{D}(\mathfrak{a}) \rightarrow \mathfrak{C}(\mathfrak{a})$ we can pass to $\mathfrak{C}(\mathfrak{a}_1)$ and inversely.
4) The formula schemata:

$$(\mathfrak{x})\,(\mathfrak{D}(\mathfrak{x}) \rightarrow \mathfrak{C}(\mathfrak{x})) \leftrightarrow (\mathfrak{z}_1)\mathfrak{C}(\mathfrak{z}_1), \quad (E\mathfrak{x})\,(\mathfrak{D}(\mathfrak{x})\ \&\ \mathfrak{C}(\mathfrak{x})) \leftrightarrow (E\mathfrak{z}_1)\mathfrak{C}(\mathfrak{z}_1).$$

5) $\{\mathfrak{x}_1 \,|\, \mathfrak{A}(\mathfrak{x}_1)\} \equiv \{\mathfrak{x} \,|\, \mathfrak{A}(\mathfrak{x})\ \&\ \mathfrak{D}(\mathfrak{x})\}$ and

$$\iota_{\mathfrak{x}_1}(\mathfrak{A}(\mathfrak{x}_1), a) = \iota_{\mathfrak{x}}((\mathfrak{A}(\mathfrak{x})\ \&\ \mathfrak{D}(\mathfrak{x})), a).$$

By the rule 3) the substitution rule yields the schema

$$\frac{\mathfrak{A}(\mathfrak{a}_1)}{\mathfrak{D}(\mathfrak{t}) \rightarrow \mathfrak{A}(\mathfrak{t})} \quad \text{as also} \quad \frac{\mathfrak{A}(\mathfrak{a}_1)}{\mathfrak{A}(\mathfrak{b}_1)},$$

and likewise every term for which $\mathfrak{D}(\mathfrak{t})$ is provable can be substituted for a new free variable. Further in virtue of 3) and 4) the inference schemata for the quantifiers with respect to the new variables and the modified formula schemata become derivable using the original schemata and the propositional calculus.

Moreover every formula or part of a formula $(\mathfrak{x})(\mathfrak{D}(\mathfrak{x}) \rightarrow \mathfrak{C}(\mathfrak{x}))$

or $(E\mathfrak{x})(\mathfrak{D}(\mathfrak{x})\ \&\ \mathfrak{C}(\mathfrak{x}))$ are replaceable by $(\mathfrak{z}_1)\mathfrak{C}(\mathfrak{z}_1)$ or $(E\mathfrak{z}_1)\mathfrak{C}(\mathfrak{z}_1)$ respectively, and inversely. The schemata for the class operator and the ι-operator with the variable c and the bound variables $\mathfrak{x}, \mathfrak{y}$, replaced by corresponding new variables become provable.

I, § 2. Equality and extensionality. Application to Descriptions

In set theory it is natural to treat together the equality and extensionality axioms. Indeed sets are regarded here as individuals which, like concrete collections, are determined by their elements.

Conforming to this idea we adopt the two axioms:

E 1 $$a=b\cdot \rightarrow \cdot a \in A \rightarrow b \in A,$$

E 2 $$(x)(x \in a \leftrightarrow x \in b) \rightarrow a=b.$$

From E 1 we immediately get, by substituting for A a class term $\{x \mid \mathfrak{A}(x)\}$ and then applying the Church schema, the equality schema

2.1 $$a=b\cdot \rightarrow \cdot \mathfrak{A}(a) \rightarrow \mathfrak{A}(b).$$

By specializations of 2.1 we get

2.2 $$a=b\cdot \rightarrow \cdot a=c \rightarrow b=c$$

2.3 $$a=b\ \&\ b=c \rightarrow a=c$$

2.4 $$a=b \rightarrow b=a.$$

The last formula results by first deriving $a=b\cdot \rightarrow \cdot a=b \rightarrow b=a$.

As an immediate consequence of E 2 we have

2.5 $$a=a,$$

so that we need not state it here separately as an axiom.

By application of 2.1 we further obtain

2.6 $$a=b\cdot \rightarrow \cdot a \in c \rightarrow b \in c$$

2.7 $$a=b\cdot \rightarrow \cdot c \in a \rightarrow c \in b.$$

From the last formula in connection with 2.4 we can derive

2.8 $$a=b \rightarrow (x)(x \in a \leftrightarrow x \in b)$$

which together with E 2 gives

2.9 $$a=b \leftrightarrow (x)(x \in a \leftrightarrow x \in b).$$

As in the derivation of these formulas the application of E 1 in most cases is by means of 2.1 in connection eventually with 2.5. Derivations of this kind will in the following be indicated briefly as "by the equality axioms".

We could have taken, as A. Fraenkel did [1926], the formula 2.9 as defining equality. Then E 2 would become derivable.

There would also be the possibility of defining equality by the equivalence

2* $$a=b \leftrightarrow (x)(a \in x \leftrightarrow b \in x)$$

what would be in the line of the definition given in *Principia Mathematica*. But the difference here is that this equivalence, as already the implication

2** $$(x)(a \in x \leftrightarrow b \in x) \rightarrow a=b,$$

is not derivable from E 1 and E 2 alone, as can be seen by simple models.

We do not introduce here equality by an explicit definition, because we want to suggest the interpretation of equality as individual identity, whereas by 2.9 or 2* taken as definiton, equality is introduced only as an equivalence relation. Besides by employing an explicit definition of equality the axioms which include equality become more complicate, if expressed with the primitive symbols.

With regard to the axiom of extensionality E 2 it is to be observed that by the unrestricted form of its statement the possibility of different "Urelemente" in the sense of Zermelo, i.e. of different individuals, which have no elements, is excluded. Thus our axiom of extensionality implies the assumption that all elements in our system are themselves sets. This – at the first glance astonishing –

feature, common to most newer axiomatic systems of set theory, can be understood as a result of an extension of Dedekind's method of introducing the real numbers, which indeed have the rôle of elements in analysis, as sets. In view of the program of embracing all classical mathematics in the system of set theory, we are induced to strengthen the said Dedekind procedure to the effect that all mathematical objects become sets, and in fact from this device considerable simplifications are arising.

One of our first applications of the equality axioms is in connection with descriptions. We have the description schemata for the symbol $\iota_{\mathfrak{x}}(\mathfrak{A}(\mathfrak{x}), a)$. For the handling with them it will be useful to introduce, but here still without an axiom, the individual symbol 0, which we shall have in our system as a primitive symbol. With this we can define

Df 2.1 $$\iota_{\mathfrak{x}}\mathfrak{A}(\mathfrak{x}) = \iota_{\mathfrak{x}}(\mathfrak{A}(\mathfrak{x}), 0).$$

Using it we obtain from our ι-schemata by applying substitution and 2.3, 2.4, the simplified ι-schemata

2.10 $$\mathfrak{A}(c) \ \& \ (\mathfrak{x})(\mathfrak{A}(\mathfrak{x}) \rightarrow \mathfrak{x} = c) \ \rightarrow \ c = \iota_{\mathfrak{x}}\mathfrak{A}(\mathfrak{x})$$

2.11 $$(E\mathfrak{x})(\mathfrak{A}(\mathfrak{x}) \ \& \ (\mathfrak{y})(\mathfrak{A}(\mathfrak{y}) \rightarrow \mathfrak{x} = \mathfrak{y}))\cdot \vee \cdot \iota_{\mathfrak{x}}(\mathfrak{A}(\mathfrak{x}) = 0.$$

From 2.11 we get by logical derivation and 2.2

2.12 $$\overline{(E\mathfrak{x})}\mathfrak{A}(\mathfrak{x}) \rightarrow \iota_{\mathfrak{x}}\mathfrak{A}(\mathfrak{x}) = 0$$

2.13 $$\mathfrak{A}(a) \ \& \ \mathfrak{A}(b) \ \& \ a \neq b \ \rightarrow \ \iota_{\mathfrak{x}}\mathfrak{A}(\mathfrak{x}) = 0.$$

From 2.10 together with 2.2, 2.4 and 2.1 we derive

2.14 $$\mathfrak{A}(c) \ \& \ (\mathfrak{x})(\mathfrak{A}(\mathfrak{x}) \rightarrow \mathfrak{x} = c)\cdot \ \rightarrow \ \cdot b = \iota_{\mathfrak{x}}\mathfrak{A}(\mathfrak{x}) \rightarrow \mathfrak{A}(b).$$

«If there is a unique $\mathfrak{x}$ such that $\mathfrak{A}(\mathfrak{x})$ and b is identical with $\iota_{\mathfrak{x}}\mathfrak{A}(\mathfrak{x})$, then $\mathfrak{A}$ holds for b.»

One might ask why we take here 2.14 instead of

$$\mathfrak{A}(c) \ \& \ (\mathfrak{x})(\mathfrak{A}(\mathfrak{x}) \rightarrow \mathfrak{x} = c)\cdot \ \rightarrow \ \cdot \mathfrak{A}(\iota_{\mathfrak{x}}\mathfrak{A}(\mathfrak{x})).$$

This is partly in order to avoid the notational complication which arises from the circumstance that in any case where $\mathfrak{a}$ in $\mathfrak{A}(\mathfrak{a})$

stands in the scope of a bound variable, we have to make some alphabetic change of bound variables in order that $\mathfrak{A}(\iota_{\mathfrak{x}}\mathfrak{A}(\mathfrak{x}))$ becomes a formula according to our conventions. Besides 2.14 for the applications is handy in so far as the ι-terms occur mostly as defining expressions for functors or individual symbols.

Let us still note as easy consequences of 2.10 and 2.11 and 2.2, 2.4, 2.5 the formulas

2.15 $$\iota_x(x=a)=a$$

2.16 $$(\mathfrak{x})(\mathfrak{A}(\mathfrak{x}) \leftrightarrow \mathfrak{B}(\mathfrak{x})) \rightarrow \iota_{\mathfrak{x}}\mathfrak{A}(\mathfrak{x})=\iota_{\mathfrak{x}}\mathfrak{B}(\mathfrak{x}).$$

Descriptions will have a main rôle for the introduction of functors by explicit definitions. In this form of definition conditioned definitions are practically included without a special convention. For instance from a definition

$$\mathfrak{f}(a)=\iota_x(\mathfrak{B}(a)\ \&\ \mathfrak{C}(a, x))$$

we can derive by means of the ι-schemata

$$\mathfrak{B}(a) \rightarrow \mathfrak{f}(a)=\iota_x\mathfrak{C}(a, x).$$

Still it might be remarked that the passage from $\iota_{\mathfrak{x}}(\mathfrak{A}(\mathfrak{x}), a)$ to $\iota_{\mathfrak{x}}\mathfrak{A}(\mathfrak{x})$ goes likewise with a specialized bound variable $\mathfrak{x}_1$ instead of $\mathfrak{x}$.

Finally we introduce here at once the *subset* and the *proper subset relation* which are definable purely logically from the primitive predicate $\in$.

Df 2.2 $$a \subseteq b \leftrightarrow (x)(x \in a \rightarrow x \in b)$$
«a is a subset of b.»

Df 2.3 $$a \subset b \leftrightarrow a \subseteq b\ \&\ \overline{b \subseteq a}.$$

The following formulas are immediately resulting:

2.17 $$a \subseteq a,\ a \subseteq b\ \&\ b \subseteq c \rightarrow a \subseteq c$$

2.18 $$\overline{a \subset a},\ a \subset b\ \&\ b \subset c \rightarrow a \subset c,\ a \subset b \rightarrow \overline{b \subset a}.$$

Further by 2.9 we have

2.19 $$a=b \leftrightarrow a \subseteq b \,\&\, b \subseteq a,$$

from which we also obtain

2.20 $$a \subseteq b \leftrightarrow a \subset b \lor a=b.$$

I, § 3. Class formalism. Class operations

Operating with classes in axiomatic set theory is not indispensable but essentially contributes to the handiness of the formalism; moreover it makes more explicit the situation which consists with regard to the rôle of the logical conceptions in mathematics. At the same time by the class formalism one way of dealing with the set-theoretic paradoxes becomes more explicit.

Let us compare this method with that of type theory. In type theory, as well known, we start from a domain of individuals, of these we consider predicates which then are regarded as constituting again a domain of individuals to which free and bound variables apply. This process then is iterated. And further from the language of predicates one passes to the language of classes. By this way a strict separation of different levels of classes comes about which causes some complications for the edifying of mathematics, as has been stressed especially by W. V. Quine.

In our system the sets are not separated as a kind of other things from the individuals, but "$\mathfrak{a} \in \mathfrak{b}$" is a relation between individuals. On the other hand something of the idea of type theory is retained in order that we are not obliged to refrain from forming assemblages which are conceptually natural as for instance that of all sets or all ordinals. Indeed, the concept of the set of all sets – provided we are allowed to apply to such a set the usual formation processes – would give a contradiction, but we can instead operate with the class φ of all sets.

This distinction between sets and classes is not a mere artifice but has its interpretation by the distinction between a set as a collection, which is a mathematical thing, and a class as an extension of a predicate, which in comparison with the mathematical things has the character of an ideal object. This point of view suggests also to regard the realm of classes not as a fixed domain of individuals but as an open universe, and the rules we shall give for class formation need not to be regarded as limiting the possible formations but as fixing a minimum of admitted processes for class formation.

In our system we bring to appear this conception of an open universe of classes, in distinction from the fixed domain of sets, by shaping the formalism of classes in a constructive way, even to the extent of avoiding at all bound class variables, whereas with regard to sets we apply the usual predicate calculus. So in our system the existential axiomatic method is joined with a constructive formalism.

By avoiding bound class variables we have also the effect that the class formation $\{x \mid \mathfrak{A}(x)\}$ is automatically predicative, i.e. not including a reference by a quantifier to the realm of classes, so that the—so to speak—type-theoretic separation of sets and classes is preserved even in the sense of the ramified type theory.

Further the conception of classes as ideal objects in distinction from the sets as proper individuals comes to appear in our system by the absence of a primitive equality relation between classes. In fact by our already mentioned definition

Df 3.1 $$A \equiv B \leftrightarrow (x)(x \in A \leftrightarrow x \in B)$$

class equality is only an expression for equal extensionality. A special axiom of extensionality for classes is therefore not needed.

Neither do we need a special axiom expressing the substitutivity of equal classes. For, every special case of the schema

3.1 $$A \equiv B \cdot \rightarrow \cdot \mathfrak{A}(A) \rightarrow \mathfrak{A}(B)$$

is derivable from Df 3.1, with the help of the predicate calculus, using also 2.16 and the explicit definitions of defined symbols eventually occurring in $\mathfrak{A}$ and $\mathfrak{B}$. The procedure is here quite the same as that one by which the general equality schema can be reduced to special equality axioms [Hilbert and Bernays 1934, I, p. 375].

The circumstance that we do not attribute a proper individuality to classes shall not hinder us to use for explanatory language the article by the familiar way in the form of description for classes.

As already observed Df 3.1 of class equality comes in particular to be applied for the definitions of class symbols. Such a definition by a formula

$$\mathfrak{K} \equiv \{\mathfrak{x} \mid \mathfrak{A}(\mathfrak{x})\}$$

passes over, by Df 3.1 and the Church schema, to

$$(\mathfrak{x})(\mathfrak{x} \in \mathfrak{K} \leftrightarrow \mathfrak{A}(\mathfrak{x})).$$

This equivalence, but written with a free variable

$$\mathfrak{a} \in \mathfrak{K} \leftrightarrow \mathfrak{A}(\mathfrak{a}),$$

can be regarded as an alternative form of the definition of $\mathfrak{K}$.

Among the formations which are afforded by the comprehension operator $\{x \mid \mathfrak{A}(x)\}$ there are in particular the well known formations of extensional logic. We have here first the familiar formations of Boolean algebra:

Df 3.2 $\quad A \subseteq B \leftrightarrow (x)(x \in A \rightarrow x \in B)$ «A is a subclass of B»

Df 3.3 $\quad \bar{A} \equiv \{x \mid x \notin A\}$ «the complement of A»

Df 3.4 $\quad A \cup B \equiv \{x \mid x \in A \vee x \in B\}$ «the union of A and B»

Df 3.5 $\quad A \cap B \equiv \{x \mid x \in A \,\&\, x \in B\}$ «the intersection of A and B»

Df 3.6 $\quad V \equiv \{x \mid x = x\}$ «the universal class»

Df 3.7 $\quad \Lambda \equiv \{x \mid x \neq x\}$ «the empty class»

By means of Λ we can express that two classes A, B are mutually exclusive, with the formula $A \cap B \equiv \Lambda$.

From the here stated definitions the familiar laws of Boolean algebra result as derivable formulas by means of the predicate calculus.

Also the extended Boolean operations can be defined, even in the following strong form

$$\bigcup_{\mathfrak{x}}(A, \mathfrak{K}(\mathfrak{x})) \equiv \{\mathfrak{z} \mid (E\mathfrak{x})(\mathfrak{x} \in A \ \& \ \mathfrak{z} \in \mathfrak{K}(\mathfrak{x}))\},$$

$$\bigcap_{\mathfrak{x}}(A, \mathfrak{K}(\mathfrak{x})) \equiv \{\mathfrak{z} \mid (\mathfrak{x})(\mathfrak{x} \in A \rightarrow \mathfrak{z} \in \mathfrak{K}(\mathfrak{x}))\},$$

and their formal laws are derivable by means of the predicate calculus. It will not be necessary here to enter in their nearer discussion. In fact we shall need only the following more special forms of extended Boolean operations:

Df 3.8 $$\bigcup A \equiv \{z \mid (Ex)(x \in A \ \& \ z \in x)\}$$
«the union (sum) of the elements of A»,

Df 3.9 $$\bigcap A \equiv \{z \mid (x)(x \in A \rightarrow z \in x)\}$$
«the intersection of the elements of A».

Extensional logic transgresses essentially Boolean algebra by the inclusion of the "relations" which constitute the extensions of predicates with more than one argument. Relations can be reduced to classes by means of the concept of the ordered pair. We shall later on introduce the ordered pair $\langle a, b\rangle$ by an explicit definition, which comes out to a kind of normation, which will be at hand by the elementary set axioms.

Here it will be sufficient to postulate the following property of the ordered pair

$$\langle a, b\rangle = \langle c, d\rangle \rightarrow a = c \ \& \ b = d$$

«Every ordered pair $\langle a, b\rangle$ uniquely determines its first member a and its second member b.»

Note that the inverse implication expressing that an ordered pair is uniquely determined by its members follows from the equality axioms (upon any explicit definition of $\langle a, b\rangle$).

Essential concepts for the theory of relations are

Df 3.10 $$\{\mathfrak{x}\mathfrak{y} \mid \mathfrak{A}(\mathfrak{x}, \mathfrak{y})\} \equiv \{\mathfrak{z} \mid (E\mathfrak{x})(E\mathfrak{y})(\mathfrak{z} = \langle \mathfrak{x}, \mathfrak{y}\rangle \ \& \ \mathfrak{A}(\mathfrak{x}, \mathfrak{y}))\}$$
«the class of pairs $\langle \mathfrak{x}, \mathfrak{y}\rangle$ such that $\mathfrak{A}(\mathfrak{x}, \mathfrak{y})$»,

Df 3.11 $\mathrm{Ps}(A) \leftrightarrow (x)(x \in A \rightarrow (Eu)(Ev)(x = \langle u, v \rangle))$
«A is a pairclass».

Df 3.12 $\Delta_1 A \equiv \{x \mid (Ey)(\langle x, y \rangle \in A)\}$,

Df 3.13 $\Delta_2 A \equiv \{y \mid (Ex)(\langle x, y \rangle \in A)\}$

If A is a pairclass, then $\Delta_1 A$ ($\Delta_2 A$) is its «domain» («converse domain»).

Df 3.14 $\breve{A} \equiv \{xy \mid \langle y, x \rangle \in A\}$

If A is a pairclass, then $\breve{A}$ is the «converse class of A».

We obviously have

3.2 $$\breve{\breve{A}} \equiv A$$

Df 3.15 $A \mid B \equiv \{xy \mid (Ez)(\langle x, z \rangle \in A \ \& \ \langle z, y \rangle \in B)\}$

If A and B are pairclasses, then $A \mid B$ is the «composition of A and B».

The composition $A \mid B$ is associative. Indeed we have

3.3 $$(A \mid B) \mid C \equiv A \mid (B \mid C)$$

so that we shall write simply $A \mid B \mid C$.

Df 3.16 $A \times B \equiv \{xy \mid x \in A \ \& \ y \in B\}$
$A \times B$ is the «cross product of A and B».

By the cross product we can express $\mathrm{Ps}(A)$, since we have

3.4 $$\mathrm{Ps}(A) \leftrightarrow A \subseteq V \times V.$$

Concerning the combination of the operations $\cup$, $\cap$, $\smile$ and $\times$ we have the following formal laws which are derivable by the predicate calculus:

3.5 $$B \times A \equiv \overbrace{A \times B},$$

3.6 $$\overbrace{A \cap B} \equiv \breve{A} \cap \breve{B},$$

3.7 $$\overbrace{A \cup B} \equiv \breve{A} \cup \breve{B},$$

3.8 $$A \times (B \cap C) \equiv (A \times B) \cap (A \times C),$$

3.9 $$A \times (B \cup C) \equiv (A \times B) \cup (A \times C).$$

I, § 4. Functionality and mappings

The theory of pairclasses contains as an essential part the theory of functionality. In fact a *function*, considered from the extensional point of view, i.e. as a Wertverlauf in the sense of Frege, is a pairclass satisfying the condition, that every element of its domain occurs in just one pair as first member. The formal definition is:

Df 4.1 $$\mathrm{Ft}(F) \leftrightarrow \mathrm{Ps}(F) \;\&\; (x)(y)(z)(\langle x, y\rangle \in F \;\&\; \langle x, z\rangle \in F \rightarrow y = z)$$ [1].

From this definition by applying the ι-schema we get

4.1 $$\mathrm{Ft}(F) \;\&\; a \in \Delta_1 F \cdot \rightarrow \cdot \langle a, b\rangle \in F \leftrightarrow \iota_x(\langle a, x\rangle \in F) = b.$$

Hence defining

Df 4.2 $$F{}^{\backprime}a = \iota_x(\langle a, x\rangle \in F)$$

we have

4.2 $$\mathrm{Ft}(F) \;\&\; a \in \Delta_1 F \cdot \rightarrow \cdot \langle a, b\rangle \in F \leftrightarrow b = F{}^{\backprime}a.$$

Thus for a function F the relation $\langle a, b\rangle \in F$ (i.e. the relation whose extension is F), can be resolved with respect to b. The interpretation of Df 4.2 is that for a function F and an element a of its domain, $F{}^{\backprime}a$ is the *value* of F for the argument a. For the case that a is not in the domain of F or different ordered pairs with first member a are in F, $F{}^{\backprime}a$ is by Df 4.2 and 2.11 equal to 0.

Generally whenever a set in dependence on certain arguments is defined by a ι-term, the definition can be practically restricted to values of the arguments of certain kinds in the sense that for other values of the arguments the ι-term is equal to 0.

In near connection with the concept of a function is that of a *one-to-one correspondence*

Df 4.3 $$\mathrm{Crs}(K) \leftrightarrow \mathrm{Ft}(K) \;\&\; \mathrm{Ft}(\breve{K}).$$

---

[1]) As variables for functions we mostly shall use the letters F, G, H, K, L, M.

We also define

Df 4.4 $$A \mathrel{\overline{\scriptstyle K}} B \leftrightarrow \mathrm{Crs}(K) \;\&\; \Delta_1(K \cap (A \times B)) \equiv A \;\& \\ \&\; \Delta_2(K \cap (A \times B)) \equiv B.$$

«A is in a one-to-one correspondence with B by means of the mapping class K.»

Note that the defining conditions on K in Df 4.4 are obviously satisfied, if $\mathrm{Crs}(K) \;\&\; \Delta_1 K \equiv A \;\&\; \Delta_2 K \equiv B$.

For the sake of convenience we have allowed in Df 4.4 the class K to be more extensive then the one-to-one correspondence between A and B itself. That this measure is not obligatory appears by the immediate provability of

4.3 $$A \mathrel{\overline{\scriptstyle K}} B \leftrightarrow A \mathrel{\overline{\scriptstyle K \cap (A \times B)}} B$$

Definition 4.4 enables us to state and derive the laws of one-to-one correspondence between classes without applying an existential quantifier for classes. This goes by explicitly indicating the mapping pairclass; by this way also the constructive character of these laws comes explicitly to appear.

We first obviously have

4.4 $$A \mathrel{\overline{\scriptstyle \{xy \,|\, x=y\}}} A$$

4.5 $$A \mathrel{\overline{\scriptstyle K}} B \rightarrow B \mathrel{\overline{\scriptstyle \breve{K}}} A$$

4.6 $$A \mathrel{\overline{\scriptstyle K}} B \;\&\; B \mathrel{\overline{\scriptstyle L}} C \rightarrow A \mathrel{\overline{\scriptstyle K|L}} C.$$

For the proofs we have to use the formulas

4.7 $$\mathrm{Crs}(\{xy \,|\, x=y\}), \quad \mathrm{Crs}(K) \rightarrow \mathrm{Crs}(\breve{K}), \quad \mathrm{Crs}(K) \;\&\; \mathrm{Crs}(L) \rightarrow \\ \rightarrow \mathrm{Crs}(K|L).$$

The last formula expresses the composibility of one-to-one correspondences.

Next we state laws which amount to a property of substitutivity of one-to-one correspondence:

4.8 $$A \mathrel{\overline{\scriptstyle K}} C \;\&\; B \mathrel{\overline{\scriptstyle L}} D \;\&\; A \cap B \equiv \Lambda \;\&\; C \cap D \equiv \Lambda \cdot\rightarrow\cdot \\ \cdot\rightarrow\cdot\; A \cup B \mathrel{\overline{\scriptstyle \mathfrak{G}}} C \cup D,$$

with $\mathfrak{C}$ being $(K \cap (A \times C)) \cup (L \cap (B \times D))$,

4.9 $$A \mathrel{\underset{K}{\overline{\to}}} C \;\&\; B \mathrel{\underset{L}{\overline{\to}}} D \rightarrow A \times B \mathrel{\underset{\mathfrak{C}}{\overline{\to}}} C \times D,$$

with $\mathfrak{C}$ being $\{xy \mid (Eu)(Ev)(x = \langle u, v\rangle \;\&\; y = \langle K`u, L`v\rangle)\}$.

Concerning the converse class we have

4.10 $$\mathrm{Ps}(A) \rightarrow A \mathrel{\underset{\mathfrak{C}}{\overline{\to}}} \breve{A}$$

and for the cross product

4.11 $$A \times B \mathrel{\underset{\mathfrak{C}}{\overline{\to}}} B \times A,$$

with $\mathfrak{C}$ both times being $\{xy \mid (Eu)(Ev)(x = \langle u, v\rangle \;\&\; y = \langle v, u\rangle)\}$, as also

4.12 $$A \times (B \times C) \mathrel{\underset{\mathfrak{C}}{\overline{\to}}} (A \times B) \times C,$$

with $\mathfrak{C}$ being $\{xy \mid (Eu)(Ev)(Ew)(x = \langle u, \langle v, w\rangle\rangle \;\&\; y = \langle\langle u, v\rangle, w\rangle)\}$.

Still the concepts concerning the relation between classes and sets have to be considered. We say that a set *represents* a class if both have the same elements, formally

Df 4.5 $$\mathrm{Rp}(A, a) \leftrightarrow (x)(x \in A \leftrightarrow x \in a)$$

«A is represented by a».

As provable formulas concerning the relation Rp we have

4.13 $$\mathrm{Rp}(\{x \mid x \in a\}, a), \quad \mathrm{Rp}(A, a) \;\&\; \mathrm{Rp}(A, b) \rightarrow a = b;$$

thus every set represents a class, and a class can be represented only by one set. But it does not result that every class is represented by a set. We even later on in our system shall state on some classes, (in particular on V), that they are not represented.

The property of a class to be represented is defined by

Df 4.6 $$\mathrm{Rp}(A) \leftrightarrow (Ex)(\mathrm{Rp}(A, x))$$ «A is represented».

We also often use the notion

Df 4.7 $$a^* \equiv \{x \mid x \in a\}$$

«a^* is the class represented by a».

Obviously we have

4.14 $$\mathrm{Rp}(A) \leftrightarrow (Ex)(x^* \equiv A).$$

Using Df 4.7 we can express the relation "a is a subset of the class B" by $a^* \subseteq B$ and likewise the relation "A is a subclass of the set b" by $A \subseteq b^*$.

Finally still here the following class formation may be mentioned

Df 4.8 $\overset{\prec}{A} \equiv \{u \mid (Ex)(Ey)(Ez)(u = \langle\langle x, y\rangle, z\rangle \mathbin{\&} \langle x, \langle y, z\rangle\rangle \in A)\}$.

«$\overset{\prec}{A}$ is the class of triplets $\langle\langle x, y\rangle, z\rangle$ such that $\langle x, \langle y, z\rangle\rangle \in A$».

We call this operation "coupling to the left". With using this operation and also the class symbol E for $\{xy \mid x \in y\}$ we have on principle the possibility of composing all class terms $\{\mathfrak{x} \mid \mathfrak{A}(\mathfrak{x})\}$ out of a small number of primary constituents. In fact it can be shown that there is a general procedure of expressing every class term which contains no constant set term, starting from the free variables occurring in it and the symbol E, by means of the operations a^*, $\bar{A}$, $A \cap B$, $A \times B$, $\Delta_1 A$, $\breve{A}$, $\overset{\prec}{A}$.

In order to give an instance we consider the process for the case of the class term

$$\{xy \mid (Ez)(\langle x, z\rangle \in a \mathbin{\&} \langle z, y\rangle \in B)\}$$

which by Df 3.15 is $a^* \mid B$.

The following preparatory steps are useful:

[1] The class $V \equiv \{x \mid x = x\}$ is expressible by $\overline{\mathsf{E} \cap \overline{\mathsf{E}}}$

[2] The class of triplets

$$\{u \mid (Ex)(Ey)(Ez)(u = \langle\langle x, y\rangle, z\rangle \mathbin{\&} \langle x, y\rangle \in a \mathbin{\&} \langle y, z\rangle \in B)\}$$

is expressible by $(((V \times V) \times V) \cap (a^* \times V)) \cap \overset{\prec}{\overline{((V \times (V \times V)) \cap (V \times B))}}$

Denoting this class by $\mathfrak{C}(a, B)$ we obtain the class to be expressed in the form $\breve{\Delta}_1(\overset{\prec}{\overset{\prec}{\mathfrak{C}}}(a, B))$.

The general proof of the stated theorem, as it results from the proof of the class theorem given in [Bernays 1937, I] and a sharpening remark in [Bernays 1954, VII], is here left for the second volume; we shall in the following not have to make use of it.

CHAPTER II

THE START OF GENERAL SET THEORY

II, § 1. The axioms of general set theory

In the formal frame described in chapt. I we now are setting up a system of general set theory – in a certain specific sense. The leading idea for it is the following: In ordinary arithmetic we have one starting element and one procedure of progressing – the successor function. In Cantor's theory of transfinite numbers we have moreover the limit process.

Now in a parallelism to these three fundamental notions we introduce three constants which we axiomatically characterize with respect to the element relation: The symbols of these constants are the individual symbol "0" – used already before without an axiomatic characterization – the binary functor "$a; b$" and the operator "$\sum_{\mathfrak{x}}(m, \mathfrak{t}(\mathfrak{x}))$" [1].

As to the last one, its formal rôle is such that for any given term $\mathfrak{t}(c)$ and a bound variable $\mathfrak{x}$ it yields a term. The characterizing axioms, more exactly two axioms and one axiom schema, for the three constants are

A 1 $$a \notin 0$$

A 2 $$a \in b; c \leftrightarrow a \in b \vee a = c$$

A 3 $$a \in \sum_{\mathfrak{x}}(m, \mathfrak{t}(\mathfrak{x})) \leftrightarrow (E\mathfrak{x})(\mathfrak{x} \in m \;\&\; a \in \mathfrak{t}(\mathfrak{x})).$$

«0 is an empty set, $b; c$ is a set whose elements are the elements of b and the set c, and $\sum_{\mathfrak{x}}(m, \mathfrak{t}(\mathfrak{x}))$ is a set whose elements are those sets which are in at least one set $\mathfrak{t}(c)$ for a c which is

---

[1]) Instead of "$\sum_{\mathfrak{x}}(m, \mathfrak{t}(\mathfrak{x}))$" we could write in a more familiar way "$\sum_{\mathfrak{x} \in m} \mathfrak{t}(\mathfrak{x})$"; only this form of expression—with "$\mathfrak{x} \in m$" as a symbolic constituent—though well suiting for communicative indication, is not fully syntactically correct for an operator in a formal system.

in m, or briefly, a set which is the union of the sets $\mathfrak{t}(c)$ such that $c \in m$.»

From the extensionality axiom it follows that 0, $b; c$ and $\sum\limits_{\mathfrak{x}}(m\ \mathfrak{t}(\mathfrak{x}))$ for given arguments are uniquely determined by the properties expressed by A 1, A 2, A 3. Indeed we have

1.1 $$(x)(x \notin c) \to c = 0,$$

1.2 $$(x)(x \in c \leftrightarrow x \in a \lor x = b) \ \to\ c = a; b,$$

1.3 $$(\mathfrak{x})(\mathfrak{x} \in c \leftrightarrow (E\mathfrak{y})(\mathfrak{y} \in m \ \&\ \mathfrak{x} \in \mathfrak{t}(\mathfrak{y}))) \ \to\ c = \sum_{\mathfrak{x}}(m, \mathfrak{t}(\mathfrak{x})).$$

Immediately connected with 1.2 and 1.3 are the following equality formulas, also resulting from A 2, A 3 by applying the axioms of equality and extensionality:

1.4 $$a = c \ \&\ b = d \to a; b = c; d,$$

1.5a $$m = n \to \sum_{\mathfrak{x}}(m, \mathfrak{s}(\mathfrak{x})) = \sum_{\mathfrak{x}}(n, \mathfrak{s}(\mathfrak{x})),$$

1.5b $$(\mathfrak{x})(\mathfrak{x} \in m \to \mathfrak{s}(\mathfrak{x}) = \mathfrak{t}(\mathfrak{x})) \to \sum_{\mathfrak{x}}(m, \mathfrak{s}(\mathfrak{x})) = \sum_{\mathfrak{x}}(m, \mathfrak{t}(\mathfrak{x})).$$

From the primitive notions 0 and $a; b$ we immediately come to the concepts of the *unit set* and the *unordered* and *ordered pair*:

Df 1.1 $$[a] = 0; a,$$

Df 1.2 $$[a, b] = [a]; b,$$

Df 1.3 $$\langle a, b \rangle = [[a], [a, b]].$$

By Df 1.1 we have

1.6 $$c \in [a] \leftrightarrow c = a,$$

and from this formula we also get

1.7 $$(x)(a \in x \leftrightarrow b \in x) \leftrightarrow a = b$$

which is the forementioned formula I, 2**.

Further we have, using A 2, 1.6 and E 2, I, 2.1,

1.8 $$c \in [a, b] \leftrightarrow c = a \lor c = b,$$

1.9 $$[a, b] = [b, a],$$

1.10 $$[a, a] = [a],$$

1.11 $$(Ex)(c \in x \mathbin{\&} x \in \langle a, b\rangle) \leftrightarrow c=a \vee c=b,$$

1.12 $$(x)(x \in \langle a, b\rangle \rightarrow c \in x) \leftrightarrow c=a$$

and from the last two formulas we get, using I, 2.1

1.13 $$\langle a, b\rangle=\langle c, d\rangle \rightarrow a=c \mathbin{\&} b=d.$$

Thus the binary functor $\langle a, b\rangle$, defined by Df 1.3, satisfies the property of the ordered pair postulated in I, § 3. Out of ordered pairs also *triplets* can be formed: $\langle a, b, c\rangle=\langle a, \langle b, c\rangle\rangle$. In the same way quadruplets, quintuplets ... and generally k-tuplets can be defined.

Now we come to state the immediate consequences of A 3. But before we want to show that the schema A 3 can be replaced, using a class variable in connection with a description, by a proper axiom. Namely from A 3 we get by taking $F'c$ for $\mathfrak{t}(c)$:

$$a \in \sum_x(m, F'x) \leftrightarrow (Ex)(x \in m \mathbin{\&} a \in F'x)$$

and defining

Df 1.4 $$\sum(m, F)=\sum_x(m, F'x)$$

we have

1.14 $$a \in \sum(m, F) \leftrightarrow (Ex)(x \in m \mathbin{\&} a \in F'x).$$

On the other hand, if we take $\sum(m, F)$ as a primitive symbol and 1.14 as an axiom and then define

$$\sum_{\mathfrak{x}}(m, \mathfrak{t}(\mathfrak{x}))=\sum(m, \{\mathfrak{u}\mathfrak{w} \mid \mathfrak{t}(\mathfrak{u})=\mathfrak{w}\})$$

we can derive every formula which is a special case of the schema A 3. Indeed substituting $\{\mathfrak{u}\mathfrak{w} \mid \mathfrak{t}(\mathfrak{u})=\mathfrak{w}\}$ for F in 1.14 we get by the above definition

[1] $$a \in \sum_{\mathfrak{x}}(m, \mathfrak{t}(\mathfrak{x})) \leftrightarrow (E\mathfrak{x})(\mathfrak{x} \in m \mathbin{\&} a \in \{\mathfrak{u}\mathfrak{w} \mid \mathfrak{t}(\mathfrak{u})=\mathfrak{w}\}'\mathfrak{x}).$$

Further by Df I, 4.2 and Df I, 3.10 we have

$$\{\mathfrak{u}\mathfrak{w} \mid \mathfrak{t}(\mathfrak{u})=\mathfrak{w}\}'c \;=\; \iota_{\mathfrak{y}}(\langle c, \mathfrak{y}\rangle \in \{\mathfrak{z} \mid (E\mathfrak{u})(E\mathfrak{w})(\mathfrak{z}=\langle \mathfrak{u}, \mathfrak{w}\rangle \mathbin{\&} \mathfrak{t}(\mathfrak{u})=\mathfrak{w})\})$$

and by the Church schema, 1.13 and the equality axioms

$$\begin{aligned}\{\mathfrak{u}\mathfrak{w} \,|\, \mathfrak{t}(\mathfrak{u})=\mathfrak{w}\}'c &= \iota_{\mathfrak{y}}(E\mathfrak{u})(E\mathfrak{w})(\langle c, \mathfrak{y}\rangle = \langle \mathfrak{u}, \mathfrak{w}\rangle \mathbin{\&} \mathfrak{t}(\mathfrak{u})=\mathfrak{w}))\\ &= \iota_{\mathfrak{y}}(E\mathfrak{u})(E\mathfrak{w})(c=\mathfrak{u} \mathbin{\&} \mathfrak{y}=\mathfrak{w} \mathbin{\&} \mathfrak{t}(\mathfrak{u})=\mathfrak{w})\\ &= \iota_{\mathfrak{y}}(\mathfrak{t}(c)=\mathfrak{y})\end{aligned}$$

which gives by I, 2.15

1.15 $$\{\mathfrak{u}\mathfrak{w} \,|\, \mathfrak{t}(\mathfrak{u})=\mathfrak{w}\}'c = \mathfrak{t}(c),$$

and combining this with [1] we come back to A 3.

Note that by 1.15 every term $\mathfrak{t}(c)$ is expressible in the form $\mathfrak{K}'c$, with $\mathrm{Ft}(\mathfrak{K})$ being provable. Indeed we have for every term $\mathfrak{t}(c)$: $\mathrm{Ft}(\{\mathfrak{u}\mathfrak{w} \,|\, (\mathfrak{t}(\mathfrak{u})=\mathfrak{w})\})$.

It may further be observed that formally 1.14 applies to an arbitrary class F and $\sum(m, F)$ is to be interpreted as the sum of those sets b for which there exists an a in m such that b is the only set such that $\langle a, b\rangle$ belongs to F. The circumstance that this dependence of $\sum(m, F)$ on F is more complicate then that of $\sum_{\mathfrak{x}}(m, \mathfrak{t}(\mathfrak{x}))$ on $\mathfrak{t}$ is the reason why we have taken A 3 and not 1.14 as axiom.

As a simple specialization of A 3 we have

1.16 $$a \in \sum_x(m, x) \leftrightarrow (Ex)(x \in m \mathbin{\&} a \in x).$$

We still define

Df 1.5 $$\sum m = \sum_x(m, x).$$

$\sum m$ is the sum of the elements of m, and 1.16 is the assertion of Zermelo's sum axiom.

Defining in particular

Df 1.6 $$a \cup b = \sum[a, b]$$

we get by 1.8 and Df 1.5

1.17 $$c \in a \cup b \leftrightarrow c \in a \vee c \in b.$$

«$a \cup b$ is the *union* (set sum) of a and b».

The formula 1.17 comes out to state that the class $a^* \cup b^*$ is represented by a set, what amounts to the same as

1.18 $$\mathrm{Rp}(A) \mathbin{\&} \mathrm{Rp}(B) \rightarrow \mathrm{Rp}(A \cup B).$$

In connection with 1.18 the question arises if not a corresponding statement holds for the intersection $A \cap B$. We shall prove even more, namely that it is sufficient for $A \cap B$ to be represented that only one of the classes A, B is represented, so that we generally have the intersection of a set a and a class B as a set. This statement comes out nearly to the assertion of Zermelo's Aussonderungs-axiom.

II, § 2. Aussonderungstheorem. Intersection

The idea of Zermelo's Aussonderungsaxiom is to admit comprehending into a set all individuals having a certain property (definite Eigenschaft) upon the condition that the individuals in question all are elements of a certain fixed set. For precising the concept of definite Eigenschaft several devices have been given. W. V. Quine [1936] observed that when the axiom system of set theory is presented as a formal system, then so to speak automatically a way of precising is given, in particular when the Aussonderungsaxiom is expressed by a schema. In our system the assertion corresponding to Zermelo's Aussonderungsaxiom does not figure as an axiom but as a provable theorem (Aussonderungs-theorem). Besides by the class formalism we are able to express this theorem by a proper formula.

2.1 $$A \subseteq B \;\&\; \mathrm{Rp}(B) \rightarrow \mathrm{Rp}(A).$$

For the proof we first eliminate the concept Rp by I, 4.14

$$A \subseteq B \;\&\; (Ex)(x^* \equiv B) \rightarrow (Ey)(y^* \equiv A).$$

By the predicate calculus this formula follows from

$$A \subseteq B \;\&\; b^* \equiv B \rightarrow (Ey)(y^* \equiv A)$$

and since by Df I, 3.2, Df I, 3.1 and Df I, 3.5 we have

$$A \subseteq B \;\&\; b^* \equiv B \rightarrow A \equiv b^* \cap A,$$

it is sufficient to prove

$$(Ey)(y^* \equiv b^* \cap A),$$

or what by Df I, 4.7 comes out to the same:

2.2 $$(Ey)(u)(u \in y \leftrightarrow u \in b \,\&\, u \in A).$$

This goes in the following way: Let $\mathfrak{C}(c, d)$ be the expression

[1] $$c \in A \rightarrow d = [c] \cdot \& \cdot c \notin A \rightarrow d = 0.$$

By the propositional calculus we have

$$c \in A \cdot\!\rightarrow\!\cdot \mathfrak{C}(c, d) \leftrightarrow d = [c]$$

and by the ι-schema I, 2.10

[2] $$c \in A \rightarrow \iota_z \mathfrak{C}(c, z) = [c].$$

Likewise we have

$$c \notin A \cdot\!\rightarrow\!\cdot \mathfrak{C}(c, d) \leftrightarrow d = 0$$
$$c \notin A \rightarrow \iota_z \mathfrak{C}(c, z) = 0.$$

Thus

[3] $$\left\{\begin{array}{lll} a \in \iota_z \mathfrak{C}(c, z) & \rightarrow c \in A & \\ & \rightarrow \iota_z \mathfrak{C}(c, z) = [c] & \text{by [2]} \\ & \rightarrow a \in [c] & \\ & \rightarrow a = c & \\ & \rightarrow a \in A. & \end{array}\right.$$

From [2] we get

[4] $$a \in b \,\&\, a \in A \rightarrow (Ex)(x \in b \,\&\, a \in \iota_z \mathfrak{C}(x, z)),$$

and from [3]

$$c \in b \,\&\, a \in \iota_z \mathfrak{C}(c, z) \rightarrow a \in b \,\&\, a \in A$$

and thus also

[5] $$(Ex)(x \in b \,\&\, a \in \iota_z \mathfrak{C}(x, z)) \rightarrow a \in b \,\&\, a \in A.$$

[4] and [5] together give

$$(Ex)(x \in b \,\&\, a \in \iota_z \mathfrak{C}(x, z)) \leftrightarrow a \in b \,\&\, a \in A.$$

Therefore by A 3 we have

$$a \in \sum_x (b, \iota_z \mathfrak{C}(x, z)) \leftrightarrow a \in b \,\&\, a \in A,$$

which gives immediately formula 2.2.

At the same time we obtain

$$\mathrm{Rp}(b^* \cap A,\ \sum_x(b,\ \iota_z\mathfrak{C}(x,z)).$$

Thus defining

Df 2.1 $\quad b \cap A = \sum_x(b,\ \iota_z(x \in A \rightarrow z=[x] \cdot \& \cdot x \notin A \rightarrow z=0))$

we have

2.3 $$a \in b \cap A \leftrightarrow a \in b \,\&\, a \in A,$$

2.4 $$\mathrm{Rp}(b^* \cap A,\ b \cap A)$$

«For any set b and any class A, $b \cap A$ is a set which is the intersection of b and A.»

For the sake of handiness we define also $A \cap b = b \cap A$.

Of 2.3 there are several direct applications. Substituting $\{x \mid \mathfrak{A}(x)\}$ for A we have by the Church schema

2.5 $$a \in b \cap \{\mathfrak{x} \mid \mathfrak{A}(\mathfrak{x})\} \leftrightarrow a \in b \,\&\, \mathfrak{A}(a).$$

This is the Aussonderungsschema. Namely by 2.5 for any set b and any predicate $\mathfrak{A}(c)$ the intersection $b \cap \{\mathfrak{x} \mid \mathfrak{A}(\mathfrak{x})\}$ is the set of all elements a of b satisfying $\mathfrak{A}(a)$. Further defining

Df 2.2 $$b \cap c = b \cap c^*$$

we have by 2.3

2.6 $$a \in b \cap c \leftrightarrow a \in b \,\&\, a \in c;$$

thus $b \cap c$ is the ordinary set intersection.

We also now can define the set difference

Df 2.3 $$b^{-}c = b \cap \overline{c^*},$$

so that we have

2.7 $$a \in b^{-}c \leftrightarrow a \in b \,\&\, a \notin c, \quad b^{-}c \subseteq b, \quad c \cap b^{-}c = 0,$$
$$b \subseteq c \rightarrow b^{-}c = 0, \quad b \subset c \rightarrow c^{-}b \neq 0.$$

By 2.6, 2.7, and 1.17 the boolean algebra for sets is available.

Moreover the binary intersection of sets can be generalized to the intersection ranging over the elements of a class. Here only we have to regard the circumstance that by Df I, 3.9 the intersection $\bigcap A$ in the case that $A \equiv \Lambda$ becomes V. This is to the effect that we can prove the class $\bigcap A$ to be represented only if A is not empty. We have by Df I, 3.9

$$c \in A \rightarrow (a \in \bigcap A \rightarrow a \in c),$$

therefore

$$c \in A \rightarrow (a \in \bigcap A \cap c \leftrightarrow a \in \bigcap A)$$

and so

2.8 $$c \in A \rightarrow \mathrm{Rp}(\bigcap A).$$

Hence, if we define

Df 2.4 $$\bigcap_{\mathfrak{x}} \mathfrak{A}(\mathfrak{x}) = \iota_{\mathfrak{z}}(\mathfrak{z}^* \equiv \bigcap\{\mathfrak{x} \mid \mathfrak{A}(\mathfrak{x})\}),$$

we have

2.9 $$\mathfrak{A}(c) \rightarrow \mathrm{Rp}(\bigcap\{\mathfrak{x} \mid \mathfrak{A}(\mathfrak{x})\}, \bigcap_{\mathfrak{x}} \mathfrak{A}(\mathfrak{x}))$$

as also

2.10 $$a \in \bigcap_{\mathfrak{x}} \mathfrak{A}(\mathfrak{x}) \quad \leftrightarrow \quad (E\mathfrak{x})\mathfrak{A}(\mathfrak{x}) \;\&\; (\mathfrak{x})(\mathfrak{A}(\mathfrak{x}) \rightarrow a \in \mathfrak{x}).$$

With respect to the generalization of the binary union of sets it is to be observed that there is no general representation of $\bigcup A$. A particular statement on representation of classes $\bigcup A$ consists by the axiom A 3, as will appear in the next considerations.

II, § 3. Sum theorem. Theorem of replacement

We have obtained among the consequences of our axioms several statements on representation of classes by sets. That is not at all astonishing. In fact, the axioms A 1–A 3 themselves are each equivalent to a statement on representation of a class. Indeed we have by using the definitions of Λ and Rp

3.1 $$(x)(x \notin 0) \leftrightarrow \mathrm{Rp}(\Lambda, 0)$$

3.2 $$(x)(x \in b; c \leftrightarrow x \in b \vee x = c) \leftrightarrow \mathrm{Rp}(\{x \mid x \in b \vee x = c\}, b; c)$$

3.3 $$(\mathfrak{z})(\mathfrak{z} \in \sum_{\mathfrak{x}}(m, \mathfrak{t}(\mathfrak{x})) \leftrightarrow (E\mathfrak{x})(\mathfrak{x} \in m \;\&\; \mathfrak{z} \in \mathfrak{t}(\mathfrak{x})) \;\cdot\!\leftrightarrow\!\cdot$$
$$\cdot\!\leftrightarrow\!\cdot\; \mathrm{Rp}(\{\mathfrak{z} \mid (E\mathfrak{x})(\mathfrak{x} \in m \;\&\; \mathfrak{z} \in \mathfrak{t}(\mathfrak{x}))\}, \sum_{\mathfrak{x}}(m, \mathfrak{t}(\mathfrak{x}))).$$

We therefore could state our axioms as existential assertions on representation:

3.4 $$(Ex)(\mathrm{Rp}(\Lambda, x) \text{ i.e. } \mathrm{Rp}(\Lambda)$$

3.5 $$\mathrm{Rp}(\{x \mid x \in b \vee x = c\})$$

3.6a $$\mathrm{Rp}(\{\mathfrak{z} \mid (E\mathfrak{x})(\mathfrak{x} \in m \; \& \; \mathfrak{z} \in \mathrm{t}(\mathfrak{x}))\}),$$

or else, corresponding to 1.14,

3.6b $$\mathrm{Rp}(\{z \mid (Ex)(x \in m \; \& \; z \in F\text{`}x)\})$$

and then introduce the symbols 0, ;, $\sum$ by explicit definitions.

The statement 3.6b can still be brought in a more handy form. Namely from 3.6b we first get by E 1

$$\Delta_1 F \equiv m^* \rightarrow \mathrm{Rp}(\{z \mid (Ex)(x \in \Delta_1 F \; \& \; z \in F\text{`}x)\})$$

and from this, since here m does not occur on the right side, by I, 4.14

[1] $$\mathrm{Rp}(\Delta_1 F) \rightarrow \mathrm{Rp}(\{z \mid (Ex)(x \in \Delta_1 F \; \& \; z \in F\text{`}x)\}).$$

Further by I, 4.2 and Df I, 3.12, Df I, 3.13 we have

$$\mathrm{Ft}(F) \cdot\!\rightarrow\!\cdot (Ex)(x \in \Delta_1 F \; \& \; b = F\text{`}x) \leftrightarrow b \in \Delta_2 F,$$

and from this we get by the predicate calculus

$$\mathrm{Ft}(F) \cdot\!\rightarrow\!\cdot (Ex)(Eu)(x \in \Delta_1 F \; \& \; c \in u \; \& \; u = F\text{`}x) \leftrightarrow$$
$$\leftrightarrow (Eu)(c \in u \; \& \; u \in \Delta_2 F).$$

Transforming the left side of the biimplication by the equality axioms and its right side by using Df I, 3.8 we obtain

[2] $$\mathrm{Ft}(F) \cdot\!\rightarrow\!\cdot (Ex)(x \in \Delta_1 F \; \& \; c \in F\text{`}x) \leftrightarrow c \in \bigcup \Delta_2 F.$$

[1] and [2] together yield

3.7 $$\mathrm{Ft}(F) \; \& \; \mathrm{Rp}(\Delta_1 F) \rightarrow \mathrm{Rp}(\bigcup \Delta_2 F).$$

«If the domain of a function is represented then the sum of its values is also represented.» (*Sum theorem*).

From the sum theorem we can go back to 3.6b and hence also to A 3. Namely denoting by $\mathfrak{K}$ the class expression

$$\{xy \mid x \in m \; \& \; y = F\text{`}x\}$$

we have

[3] $\quad \mathrm{Ft}(\mathfrak{K}), \quad \Delta_1\mathfrak{K} \equiv m^*, \quad (x)(x \in m \to \mathfrak{K}`x = F`x).$

From the first two formulas we get by 3.7: $\mathrm{Rp}(\bigcup\Delta_2\mathfrak{K})$ and from this by [2], what was derived without using A 3,

$$\mathrm{Rp}(\{z \mid (Ex)(x \in \Delta_1\mathfrak{K} \;\&\; z \in \mathfrak{K}`x)\}).$$

But, by the second and the third of the formulas [3], this yields

$$\mathrm{Rp}(\{z \mid (Ex)(x \in m \;\&\; z \in F`x)\}).$$

From the sum theorem we pass to the *theorem of replacement.* Indeed denoting by $\mathfrak{K}$ the class term

$$\{xy \mid (Eu)(\langle x, u\rangle \in F \;\&\; y = [u])\})$$

we have

$$\mathrm{Ft}(F) \to \mathrm{Ft}(\mathfrak{K}),\ \Delta_1\mathfrak{K} \equiv \Delta_1 F,\ \bigcup\Delta_2\mathfrak{K} \equiv \Delta_2 F.$$

Therefore by 3.7 we have

3.8 $\quad \mathrm{Ft}(F) \;\&\; \mathrm{Rp}(\Delta_1 F) \to \mathrm{Rp}(\Delta_2 F).$

«If the domain of a function is represented then also the class of its values.»

The content of 3.8 is the assertion of Fraenkel's axiom of replacement.

Concerning the proof of 3.8 we observe that the class $\bigcup\Delta_2\mathfrak{K}$ there occurring is nothing else than the class represented by

$$\sum_x(m, [F`x])$$

with $m^* \equiv \Delta_1 F$. And indeed the proof of 3.8 can also be given by stating the formula

$$\mathrm{Ft}(F) \;\&\; \Delta_1 F \equiv m^* \to \mathrm{Rp}(\Delta_2 F, \sum_x(m, [F`x]).$$

For the application of the replacement theorem it will be sometimes useful to have at hand the following *replacement operator*:

Df 3.1 $\quad \prod_{\mathfrak{x}}(m, \mathfrak{t}(\mathfrak{x})) = \sum_{\mathfrak{x}}(m, [\mathfrak{t}(\mathfrak{x})]).$

We have

3.9 $\quad a \in \prod_{\mathfrak{x}}(m, \mathfrak{t}(\mathfrak{x})) \leftrightarrow (E)(\mathfrak{x})(\mathfrak{x} \in m \;\&\; a = \mathfrak{t}(\mathfrak{x})).$

Let us consider the next consequences of 3.8. We have first

3.10 $$\mathrm{Ft}(F) \,\&\, \mathrm{Rp}(\Delta_1 F) \rightarrow \mathrm{Rp}(F).$$

For, denoting by $\mathfrak{R}$ the class term

$$\{xy \,\|\, (Ez)(\langle x, z\rangle \in F \,\&\, y = \langle x, z\rangle)\})$$

we have

$$\mathrm{Ft}(F) \rightarrow \mathrm{Ft}(\mathfrak{R}),\ \Delta_1\mathfrak{R} \equiv \Delta_1 F,\ \Delta_2\mathfrak{R} \equiv F,$$

what by 3.8 gives 3.10.

A further application of 3.8 is to arbitrary pairclasses. If A is any pairclass then the class

$$\{xy \,\|\, (Ez)(x = \langle y, z\rangle \,\&\, x \in A)\})$$

is a function whose domain is A und whose converse domain is $\Delta_1 A$. Therefore by 3.8 we have

3.11 $$\mathrm{Ps}(A) \,\&\, \mathrm{Rp}(A) \rightarrow \mathrm{Rp}(\Delta_1 A).$$

In quite a corresponding way we obtain

3.12 $$\mathrm{Ps}(A) \,\&\, \mathrm{Rp}(A) \rightarrow \mathrm{Rp}(\Delta_2 A).$$

By our definitions of $\Delta_1 A$ and $\Delta_2 A$ it would even be possible to omit the premise $\mathrm{Ps}(A)$ in 3.11 and 3.12. For, if A is an arbitrary class which is represented, then by the Aussonderungstheorem the class of pairs belonging to A is also represented.

In order to prove the inverse of 3.11 and 3.12 i.e.

3.13 $$\mathrm{Ps}(A) \,\&\, \mathrm{Rp}(\Delta_1 A) \,\&\, \mathrm{Rp}(\Delta_2 A) \rightarrow \mathrm{Rp}(A)$$

it will be sufficient in virtue of the Aussonderungstheorem to prove

3.14 $$\mathrm{Rp}(A) \,\&\, \mathrm{Rp}(B) \rightarrow \mathrm{Rp}(A \times B).$$

Indeed we have

$$\mathrm{Ps}(A) \quad \leftrightarrow \quad A \subseteq \Delta_1 A \times \Delta_2 A.$$

Now the proof of 3.14 goes by defining

Df 3.2 $$a \times b = \sum_x (a, \prod_y (b, \langle x, y\rangle)).$$

This immediately gives, using A 3 and 3.9,

3.15 $$\mathrm{Rp}(A, a) \,\&\, \mathrm{Rp}(B, b) \rightarrow \mathrm{Rp}(A \times B, a \times b),$$

from which 3.14 directly results. At the same time 3.15 immediately yields

3.16 $$a^* \times b^* \equiv (a \times b)^*.$$

II, § 4. Functional sets. One-to-one correspondences

By the stated facts of representation there is the possibility of carrying over many class concepts to set concepts. To make the parallelism more explicit, we shall take the liberty of using in the case where a set predicate or set function is defined by the corresponding predicate or function of the class represented by the set, the same predicate or function symbol, what indeed gives no confusion, since the symbols in question have occurred before only with class arguments.

Df 4.1 $$\mathrm{Ps}(a) \leftrightarrow \mathrm{Ps}(a^*),$$
«a is a pairset.»

Df 4.2 $$\mathrm{Ft}(f) \leftrightarrow \mathrm{Ft}(f^*),$$
«f is a functional set», or briefly «a function».

Df 4.3 $$\mathrm{Crs}(f) \leftrightarrow \mathrm{Crs}(f^*)$$

Df 4.4 $$f\text{`}a = f^*\text{`}a.$$

Whenever f is a functional set and a an element of its domain, then $f\text{`}a$ is the value of f for a, otherwise it is 0.

Df 4.5a $$\Delta_1 a = \iota_x(x^* \equiv \Delta_1(a^*)),$$

Df 4.5b $$\Delta_2 a = \iota_x(x^* \equiv \Delta_2(a^*)),$$

Df 4.6 $$a \mid b = \iota_x(x^* \equiv a^* \mid b^*),$$

Df 4.7 $$\breve{a} = \iota_x(x^* \equiv \breve{a^*}).$$

Note that we have

4.1a $$c \in \Delta_1 a \leftrightarrow (Ey)(\langle c, y\rangle \in a)$$

4.1b $$c \in \Delta_2 a \leftrightarrow (Ex)(\langle x, c\rangle \in a)$$

4.2 $\quad c \in a \mid b \leftrightarrow (Ex)(Ey)(Ez)(c = \langle x, y\rangle \;\&\; \langle x, z\rangle \in a \;\&\; \langle z, y\rangle \in b),$

4.3 $\quad c \in \breve{a} \leftrightarrow (Ex)(Ey)(c = \langle x, y\rangle \;\&\; \langle y, x\rangle \in a).$

Concerning the operations $\times$, $\smile$ composed with another and with $\cup$ and $\cap$ the laws stated for classes I, 3.5–I, 3.9 pass immediately over to the corresponding laws for sets. Indeed each of the said operations commutes with the * operation.

An essential application of 3.16 is to the theory of one-to-one correspondences between sets. We define it in the familiar way by

Df 4.8 $\quad a \sim b \leftrightarrow (Ex)(\mathrm{Crs}(x) \;\&\; \Delta_1 x = a \;\&\; \Delta_2 x = b).$

In virtue of Df I, 4.4 we have the formula

4.4 $$a \sim b \leftrightarrow (Ex)(a^* \overline{\scriptstyle x^*}\, b^*).$$

For a fully satisfactory connection between the concepts $A \overline{\scriptstyle K} B$ and $a \sim b$ it is still be required that the following formula hold

4.5 $$a^* \overline{\scriptstyle K}\, b^* \rightarrow (Ex)(a^* \overline{\scriptstyle x^*}\, b^*),$$

so that

4.6 $$a^* \overline{\scriptstyle K}\, b^* \rightarrow a \sim b.$$

4.5 can be proved as follows. We first have by I, 4.3 and 3.16

$$a^* \overline{\scriptstyle K}\, b^* \leftrightarrow a^* \overline{\scriptstyle K \cap (a^* \times b^*)}\, b^*$$
$$\leftrightarrow a^* \overline{\scriptstyle K \cap (a \times b)^*}\, b^*.$$

Now by the Aussonderungstheorem 2.4

$$\mathrm{Rp}(K \cap (a \times b)^*,\ K \cap (a \times b))$$

and so 4.5 results.

By means of 4.5 and 4.6 we are now able to derive the laws of one-to-one correspondences between sets from the corresponding laws about the class relation $A \overline{\scriptstyle K} B$. Indeed in this way from I, 4.4, I, 4.5, and I, 4.6 we get the formal laws of $a \sim b$:

4.7 $\quad a \sim a, \quad a \sim b \rightarrow b \sim a, \quad a \sim b \;\&\; b \sim c \rightarrow a \sim c,$

and from I, 4.8, I, 4.9 the laws of substitutivity

4.8 $a \sim c \;\&\; b \sim d \;\&\; a \cap b = 0 \;\&\; c \cap d = 0 \;\rightarrow\; a \cup b \sim c \cup d$

4.9 $$a \sim c \;\&\; b \sim d \rightarrow a \times b \sim c \times d.$$

Concerning $\breve{a}$ we have from I, 4.10

4.10 $$\mathrm{Ps}(a) \rightarrow a \sim \breve{a}$$

and for the crossproduct from I, 4.11 and I, 4.12

4.11 $$a \times b \sim b \times a,,$$

4.12 $$a \times (b \times c) \sim (a \times b) \times c.$$

We still here have to consider the concept of the *Belegungsklasse*, in Cantor's sense of Belegung, wherein class concepts and set concepts occur combined.

Df 4.9 $$A^{\underline{c}} \equiv \{z \mid Ft(z) \;\&\; \Delta_1 z = c \;\&\; (\Delta_2 z)^* \subseteq A)\}.$$

> «$A^{\underline{c}}$ is the class of functional sets whose domain is c and whose converse domain is a subset of A, thus it is the class of mappings (by functional sets) of c into A.»

Concerning one-to-one correspondences between Belegungsklassen we have first

4.13 $$A \mathrel{\overline{\scriptstyle K}} B \;\&\; c^* \mathrel{\overline{\scriptstyle L}} d^* \rightarrow A^{\underline{c}} \mathrel{\overline{\scriptstyle \mathfrak{C}}} B^{\underline{d}},$$

with $\mathfrak{C}$ being

$$\{uv \mid u \in A^{\underline{c}} \;\&\; \mathrm{Ft}(v) \;\&\; \Delta_1 v = d \;\&\; (x)(x \in d \rightarrow v^{\iota}x = K^{\iota}(u^{\iota}(\breve{L}^{\iota}x)))\},$$

and further the formulas which are analogous to the computation laws for the numeral power function:

4.14 $$A^{\underline{c}} \times B^{\underline{c}} \mathrel{\overline{\scriptstyle \mathfrak{C}}} (A \times B)^{\underline{c}}$$

with $\mathfrak{C}$ being $\{wz \mid (Eu)(Ev)(\mathrm{Ft}(u) \;\&\; \mathrm{Ft}(v) \;\&\; \Delta_1 u = c \;\&\; \Delta_1 v = c \;\&$
$\& \; w = \langle u, v \rangle \;\&\; z^* \equiv \{xy \mid x \in c \;\&\; y = \langle u^{\iota}x, v^{\iota}x \rangle\})\},$

4.15 $$b \cap c = 0 \rightarrow A^{\underline{b}} \times A^{\underline{c}} \mathrel{\overline{\scriptstyle \mathfrak{C}}} A^{\underline{b \cup c}}$$

with $\mathfrak{C}$ being $\{xy \mid (Eu)(Ev)(\mathrm{Ft}(u)\ \&\ \mathrm{Ft}(v)\ \&\ \Delta_1 u = b\ \&\ \Delta_1 v = c\ \&$

$\&\ x = \langle u, v\rangle\ \&\ y = u \cup v)\}$,

4.16 $$(A^{\underline{b}})^{\underline{c}} \underset{\mathfrak{C}}{\sim} A^{\underline{b \times c}}$$

with $\mathfrak{C}$ being $\{uv \mid \mathrm{Ft}(u)\ \&\ (z)(z \in c \rightarrow \mathrm{Ft}(u`z)\ \&\ \Delta_1(u`z) = b)\ \&\ \mathrm{Ft}(v)\ \&$

$\&\ \Delta_1 v = b \times c\ \&\ (x)(y)(x \in b\ \&\ y \in c \rightarrow v`\langle x, y\rangle = (u`y)`x)\}$.

The corresponding set formulas, using $a \sim b$ and 4.5, 4.6, are not yet available, since we are not able to prove in general $\mathrm{Rp}((a^*)^{\underline{b}})$.

CHAPTER III

ORDINALS; NATURAL NUMBERS; FINITE SETS

III, § 1. Fundaments of the theory of ordinals

For the edification of classical mathematics in our set theoretic and class theoretic frame it is advantageous to begin with the theory of ordinal numbers (ordinals). This theory, as was discovered independently by Zermelo and von Neumann [1923], can be set up without introducing before the concept of order type. We can define ordinals even without referring to order at all. The possibility of such a definition is given by Cantor's theorem that the set of the ordinals lower than an ordinal l has by the natural order the order type l. Our procedure now will come out to identify the ordinal l with the set of the lower ordinals, so that the ordinals become wellordered [1]) by the relation $\in$.

We begin with introducing three auxiliary concepts:

Df 1.1 $\quad \mathrm{Trans}(d) \leftrightarrow (x)(y)(x \in y \,\&\, y \in d \rightarrow x \in d).$
«d is transitive».

We immediately get by Df I, 2.2

1.1 $\quad \mathrm{Trans}(d) \leftrightarrow (y)(y \in d \rightarrow y \subseteq d).$

Df 1.2 $\quad \mathrm{Alt}(d) \leftrightarrow (x)(y)(x \in d \,\&\, y \in d \,\&\, x \neq y \rightarrow x \in y \vee y \in x)$

Df 1.3 $\quad \mathrm{Fund}(d) \leftrightarrow (x)(x \subseteq d \,\&\, x \neq 0 \rightarrow (Ey)(y \in x \,\&\, y \cap x = 0))$ [2]).

Now we define the concept of an ordinal as follows [3])

Df 1.4 $\quad \mathrm{Od}(d) \leftrightarrow \mathrm{Trans}(d) \,\&\, \mathrm{Alt}(d) \,\&\, \mathrm{Fund}(d)$

---

[1]) We are speaking of wellorder in this chapter only in the familiar intuitive sense. Later on we shall state a formal definition of wellorder (chapt. V).

[2]) Fund(d) says that d is "wohlfundiert" by the relation $\in$ in the sense of Zermelo [1935].

[3]) The method of starting from this definition of ordinal is due to R. M. Robinson [1937].

Our first task now is to show that the elements of an ordinal are wellordered by the $\in$-relation. For this it will suffice to show that the following formulas hold:

[1] $$\mathrm{Od}(d) \mathbin{\&} a \in d \rightarrow a \notin a,$$

[2] $$\mathrm{Od}(d) \mathbin{\&} a \in d \mathbin{\&} b \in d \mathbin{\&} c \in d \cdot\rightarrow\cdot a \in b \mathbin{\&} b \in c \rightarrow c \notin a \mathbin{\&} c \neq a,$$

[3] $$\mathrm{Od}(d) \rightarrow (x)(x \subseteq d \mathbin{\&} x \neq 0 \rightarrow (Ey)(y \in x \mathbin{\&} (z)(z \in x \rightarrow \rightarrow y = z \vee y \in z))).$$

Namely [1] and [2] together with $\mathrm{Od}(d) \rightarrow \mathrm{Alt}(d)$ express, that the elements of an ordinal are ordered by the relation $\in$, and [3] says that by this order every non empty subset of an ordinal has a first element.

In fact we shall prove the following stronger formulas:

1.2 $$\mathrm{Fund}(d) \mathbin{\&} a \in d \rightarrow a \notin a$$

1.3 $$\mathrm{Fund}(d) \mathbin{\&} a \in d \mathbin{\&} b \in d \mathbin{\&} c \in d \cdot\rightarrow\cdot a \in b \mathbin{\&} b \in c \rightarrow c \notin a \mathbin{\&} c \neq a$$

1.4 $$\mathrm{Fund}(d) \mathbin{\&} \mathrm{Alt}(d) \cdot\rightarrow\cdot (x)(x \subseteq d \mathbin{\&} x \neq 0 \rightarrow (Ey)(y \in x \mathbin{\&} \mathbin{\&} (z)(z \in x \rightarrow y = z \vee y \in z))).$$

These formulas together yield the statement that every set satisfying only the conditions Fund and Alt, is wellordered by the $\in$-relation. This fact was observed by Zermelo [1935]. The proof of 1.2 goes by applying the definition of Fund to the non empty subset $[a]$ of d, so that we get

$$\begin{aligned} \mathrm{Fund}(d) \mathbin{\&} a \in d &\rightarrow (Ey)(y \in [a] \mathbin{\&} y \cap [a] = 0) \\ &\rightarrow a \cap [a] = 0 \\ &\rightarrow a \notin a. \end{aligned}$$

By the same method we prove

[1] $$\mathrm{Fund}(d) \mathbin{\&} a \in d \mathbin{\&} b \in d \mathbin{\&} a \in b \rightarrow b \notin a$$

[2] $$\mathrm{Fund}(d) \mathbin{\&} a \in d \mathbin{\&} b \in d \mathbin{\&} c \in d \mathbin{\&} a \in b \mathbin{\&} b \in c \rightarrow c \notin a.$$

For the first we have to consider the subset $[a, b]$ of d, for the second the subset $[a, b]; c$, for which we have

$$k \in [a, b];\ c \rightarrow k = a \vee k = b \vee k = c.$$

From [1], [2] we obtain 1.3, and by Df 1.2 and Df 1.3 follows directly 1.4.

The next is to prove that each element of an ordinal is itself an ordinal.

1.5 $$\mathrm{Od}(d)\ \&\ c \in d \rightarrow \mathrm{Od}(c).$$

The proof goes by combining 1.3 with Df 1.1 and Df 1.4, what gives

$$\mathrm{Od}(d)\ \&\ c \in d\ \&\ b \in c\ \&\ a \in b \rightarrow a \in c$$

and thus

$$\mathrm{Od}(d)\ \&\ c \in d \rightarrow \mathrm{Trans}(c),$$

as also with 1.1

[1] $$\mathrm{Od}(d)\ \&\ c \in d \rightarrow \mathrm{Trans}(c)\ \&\ c \subseteq d.$$

Further using the obvious formulas

$$\mathrm{Alt}(d)\ \&\ c \subseteq d \rightarrow \mathrm{Alt}(c)$$
$$\mathrm{Fund}(d)\ \&\ c \subseteq d \rightarrow \mathrm{Fund}(c)$$

we get

[2] $$\mathrm{Od}(d)\ \&\ c \subseteq d\ \&\ \mathrm{Trans}(c) \rightarrow \mathrm{Od}(c).$$

[1] and [2] together give 1.5.

The formula [1] can be sharpened, using 1.2 and Df 1.4 to

1.6 $$\mathrm{Od}(d)\ \&\ c \in d \rightarrow \mathrm{Trans}(c)\ \&\ c \subset d.$$

Of formula 1.6 also the inverse holds:

1.7 $$\mathrm{Od}(d)\ \&\ c \subset d\ \&\ \mathrm{Trans}(c) \rightarrow c \in d.$$

«Every transitive proper subset of an ordinal is an element of it.»

The proof goes by applying essentially the Aussonderungstheorem. By II, 2.7 and Df 1.3 we have

[1] $$\mathrm{Fund}(d)\ \&\ c \subset d \rightarrow (Ey)(y \in d^{-}c\ \&\ y \cap d^{-}c = 0).$$

Further we have

$$\mathrm{Trans}(c)\ \&\ a \notin c\ \&\ b \in c \rightarrow a \neq b\ \&\ a \notin b$$

and therefore

$$\mathrm{Alt}(d)\ \&\ c \subset d\ \&\ \mathrm{Trans}(c)\ \&\ a \in d^{-}c\ \&\ b \in c \rightarrow b \in a,$$

what gives by the predicate calculus and Df I, 2.2

$$\mathrm{Alt}(d) \mathbin{\&} c \subset d \mathbin{\&} \mathrm{Trans}(c) \mathbin{\&} a \in d^{-}c \rightarrow c \subseteq a.$$

On the other hand we have, using 1.1,

$$\mathrm{Trans}(d) \mathbin{\&} a \in d^{-}c \mathbin{\&} a \cap d^{-}c = 0 \rightarrow a \subseteq c;$$

so we get

[2] $$\mathrm{Trans}(d) \mathbin{\&} \mathrm{Alt}(d) \mathbin{\&} c \subset d \mathbin{\&} \mathrm{Trans}(c) \mathbin{\&} a \in d^{-}c \mathbin{\&} \\ \mathbin{\&} a \cap d^{-}c = 0 \rightarrow c = a \mathbin{\&} c \in d.$$

[1] and [2] together yield by the predicate calculus and Df 1.4 the formula 1.7.

By 1.5, 1.6 and 1.7 we obtain

1.8 $$\mathrm{Od}(b) \rightarrow (a \in b \leftrightarrow \mathrm{Od}(a) \mathbin{\&} a \subset b).$$

«The elements of an ordinal b are those ordinals which are proper subsets of b.»

Now we are to show that of any two ordinals a, b one is a subset of the other.

1.9 $$\mathrm{Od}(a) \mathbin{\&} \mathrm{Od}(b) \rightarrow a \subseteq b \vee b \subseteq a.$$

By II, Df 2.2, the proof of 1.9 comes out to derive

[1] $$\mathrm{Od}(a) \mathbin{\&} \mathrm{Od}(b) \rightarrow a \cap b = a \vee a \cap b = b.$$

From Df 1.1 we immediately get

$$\mathrm{Trans}(a) \mathbin{\&} \mathrm{Trans}(b) \rightarrow \mathrm{Trans}(a \cap b).$$

Therefore we have

$$\mathrm{Od}(a) \mathbin{\&} \mathrm{Od}(b) \mathbin{\&} a \cap b \neq a \rightarrow \mathrm{Trans}(a \cap b) \mathbin{\&} a \cap b \subset a$$

and by 1.7

[2] $$\mathrm{Od}(a) \mathbin{\&} \mathrm{Od}(b) \mathbin{\&} a \cap b \neq a \rightarrow a \cap b \in a.$$

In the same way we get

[3] $$\mathrm{Od}(a) \mathbin{\&} \mathrm{Od}(b) \mathbin{\&} a \cap b \neq b \rightarrow a \cap b \in b.$$

On the other hand we have

[4] $$a \cap b \in a \mathbin{\&} a \cap b \in b \rightarrow a \cap b \in a \cap b$$

and by 1.2

$$[5] \qquad \mathrm{Od}(a) \mathbin{\&} a \cap b \in a \rightarrow a \cap b \notin a \cap b.$$

From [2]–[5] we get by the propositional calculus the formula [1].

Now 1.9 immediately gives

$$1.10 \qquad \mathrm{Od}(a) \mathbin{\&} \mathrm{Od}(b) \mathbin{\&} a \neq b \rightarrow a \subset b \vee b \subset a.$$

Thus in virtue of I, 2.18 the ordinals are ordered by the relation $\subset$. But by 1.8, for ordinals a, b the proper subset relation is equivalent with the element relation. Therefore the ordinals are also ordered by the relation $\in$, and we have

$$\mathrm{Od}(a) \mathbin{\&} \mathrm{Od}(b) \cdot\!\rightarrow\!\cdot a = b \vee a \in b \vee b \in a$$

$$1.11 \qquad \mathrm{Od}(a) \mathbin{\&} \mathrm{Od}(b) \mathbin{\&} \mathrm{Od}(c) \cdot\!\rightarrow\!\cdot a \in b \mathbin{\&} b \in c \rightarrow a \in c$$

$$\mathrm{Od}(a) \mathbin{\&} \mathrm{Od}(b) \mathbin{\&} a \in b \rightarrow b \notin a.$$

We can now speak as usual of "higher" and "lower" ordinals in the sense that an ordinal a is lower than an ordinal b, and b higher than a if $a \in b$ or if $a \subset b$. At the same time by 1.8 it results that an ordinal a is just the set of the ordinals lower than a.

Moreover we now also show that the order constituted for ordinals by the relation $\in$ is a *wellorder*. In fact we have the provable formula

$$1.12 \qquad \mathrm{Od}(c) \mathbin{\&} c \in A \rightarrow (Ex)(\mathrm{Od}(x) \mathbin{\&} x \in A \mathbin{\&} (z)(\mathrm{Od}(z) \mathbin{\&} \\ \mathbin{\&} z \in A \rightarrow x = z \vee x \in z)).$$

«Among the ordinals which are elements of a class A there is a lowest one – or what comes out to the same: every non empty class of ordinals has a lowest element.»

This is the *principle of the least ordinal number*. The proof goes as follows. By 1.4 we have

$$\mathrm{Od}(c) \cdot\!\rightarrow\!\cdot a \subseteq c \mathbin{\&} a \neq 0 \rightarrow (Ey)(y \in a \mathbin{\&} (z)(z \in a \rightarrow y = z \vee y \in z)).$$

Substituting here $c \cap A$ for a we get

$$[1] \qquad \mathrm{Od}(c) \mathbin{\&} c \cap A \neq 0 \rightarrow (Ex)(x \in c \mathbin{\&} x \in A \mathbin{\&} \\ \mathbin{\&} (z)(z \in c \mathbin{\&} z \in A \rightarrow x = z \vee x \in z)).$$

On the other hand we have by 1.11

$$[2] \qquad \mathrm{Od}(c) \rightarrow (x)(z)(x \in c \mathbin{\&} z \notin c \mathbin{\&} \mathrm{Od}(z) \rightarrow x \in z).$$

[1] and [2] together give by the predicate calculus, with 1.5,

$$\mathrm{Od}(c) \mathbin{\&} c \cap A \neq 0 \;\rightarrow\; (Ex)(\mathrm{Od}(x) \mathbin{\&} x \in A \mathbin{\&} \\ \mathbin{\&} (z)(\mathrm{Od}(z) \mathbin{\&} z \in A \;\rightarrow\; x = z \vee x \in z)).$$

Now in order to get 1.12 it is sufficient to derive

$$\mathrm{Od}(c) \mathbin{\&} c \in A \mathbin{\&} c \cap A = 0 \;\rightarrow\; (Ex)(\mathrm{Od}(x) \mathbin{\&} x \in A \mathbin{\&} \\ \mathbin{\&} (z)(\mathrm{Od}(z) \mathbin{\&} z \in A \;\rightarrow\; x = z \vee x \in z)).$$

But this results from the formula

$$\mathrm{Od}(c) \mathbin{\&} c \cap A = 0 \;\rightarrow\; (z)(\mathrm{Od}(z) \mathbin{\&} z \in A \;\rightarrow\; c = z \vee c \in z)$$

which we get by 1.11.

The formula 1.12 expresses the wellorder property of the relation $\in$ between ordinals by a generality on classes (not merely on sets), what is in conformity with the intuitive concept of wellorder. In the formulation 1.4 of the property of an ordinal that its elements are wellordered by $\in$, a generality on classes was not needed, since every class of elements of an ordinal, by the Aussonderungs-theorem, is represented by a set.

We have proved 1.12 from the property Fund of ordinals. On the other hand from 1.12 we easily obtain the consequence

$$(x)(x \in m \rightarrow \mathrm{Od}(x)) \;\rightarrow\; \mathrm{Fund}(m).$$

Since further we directly infer from 1.11 that every set of ordinals has the property Alt, we get by Df 1.4 the theorem

1.13 $\quad (x)(x \in m \rightarrow \mathrm{Od}(x)) \mathbin{\&} \mathrm{Trans}(m) \rightarrow \mathrm{Od}(m).$

«Every transitive set of ordinals is itself an ordinal».

For most of the applications of 1.12 it is useful to have at hand the notion of the *least ordinal belonging to A*. We define

Df 1.5 $\quad \mu A = \iota_x(\mathrm{Od}(x) \mathbin{\&} x \in A \mathbin{\&} (z)(\mathrm{Od}(z) \mathbin{\&} z \in A \cdot\!\rightarrow\!\cdot x = z \vee x \in z).$

The characterizing properties of μA are

1.14 $\quad \mathrm{Od}(c) \mathbin{\&} c \in A \rightarrow \mathrm{Od}(\mu A) \mathbin{\&} \mu A \in A$

1.15 $\quad \mathrm{Od}(c) \mathbin{\&} c \in A \rightarrow \mu A = c \vee \mu A \in c$

1.16 $\quad (x)(\mathrm{Od}(x) \rightarrow x \notin A) \rightarrow \mu A = 0.$

For the proof of these formulas which of course is by the ι-schema, we have mainly to use 1.12, but besides also the unicity formula

$$\mathrm{Od}(a) \mathbin{\&} a \in A \mathbin{\&} (z)(\mathrm{Od}(z) \mathbin{\&} z \in A \to a=z \vee a \in z) \mathbin{\&}$$
$$\mathbin{\&} \mathrm{Od}(b) \mathbin{\&} b \in A \mathbin{\&} (z)(\mathrm{Od}(z) \mathbin{\&} z \in A \to b=z \vee b \in z) \to a=b,$$

which results from 1.11 by the predicate calculus and the equality axioms. For a handy notation we still define

Df 1.6 $$\mu_{\mathfrak{x}}\mathfrak{A}(\mathfrak{x}) = \mu\{\mathfrak{x} \mid \mathfrak{A}(\mathfrak{x})\}.$$

From the principle of the least ordinal number we can pass to the *principle of transfinite induction* in the following way: Applying 1.12 to a class $\{\mathfrak{x} \mid \overline{\mathfrak{A}}(\mathfrak{x})\}$ and using

$$\mathrm{Od}(a) \mathbin{\&} (c=a \vee c \in a) \to a \notin c,$$

which gives

$$(\mathfrak{z})(\mathrm{Od}(\mathfrak{z}) \mathbin{\&} \overline{\mathfrak{A}}(\mathfrak{z}) \to c=\mathfrak{z} \vee c \in \mathfrak{z}) \to (\mathfrak{z})(\mathrm{Od}(\mathfrak{z}) \mathbin{\&} \mathfrak{z} \in c \to \mathfrak{A}(\mathfrak{z})),$$

we obtain by the predicate calculus, using also 1.5,

$$\mathrm{Od}(c) \mathbin{\&} \overline{\mathfrak{A}}(c) \to (E\mathfrak{x})(\mathrm{Od}(\mathfrak{x}) \mathbin{\&} \overline{\mathfrak{A}}(\mathfrak{x}) \mathbin{\&} (\mathfrak{z})(\mathfrak{z} \in \mathfrak{x} \to \mathfrak{A}(\mathfrak{z}))),$$

and from this by contraposition the formula schema of transfinite induction results:

1.17 $$(\mathfrak{x})(\mathrm{Od}(\mathfrak{x}) \mathbin{\&} (\mathfrak{z})(\mathfrak{z} \in \mathfrak{x} \to \mathfrak{A}(\mathfrak{z})) \to \mathfrak{A}(\mathfrak{x})) \cdot\!\to\!\cdot \mathrm{Od}(c) \to \mathfrak{A}(c).$$

«If, for every ordinal $\mathfrak{x}$, $\mathfrak{A}$ holds provided that it holds for every ordinal lower than $\mathfrak{x}$, then $\mathfrak{A}$ holds for every ordinal».

From 1.17 we can also go back to 1.12 by taking for $\mathfrak{A}(c)$ in 1.17 $c \notin A$ and using 1.11.

III, § 2. Existential statements on ordinals. Limit numbers

To the proved hypothetical laws concerning ordinal numbers now have to be added the existential theorems. These are in strict relatedness to our three fundamental notions 0 , $;$, $\sum$ and to the corresponding axioms A 1, A 2, A 3.

Namely we have

2.1 $$\mathrm{Od}(0)$$

2.2 $$\mathrm{Od}(c) \to \mathrm{Od}(c;c)$$

2.3 $$(\mathfrak{x})(\mathfrak{x} \in m \to \mathrm{Od}(\mathfrak{t}(\mathfrak{x}))) \to \mathrm{Od}(\sum_{\mathfrak{x}}(m, \mathfrak{t}(\mathfrak{x}))).$$

Indeed the properties Trans, Alt and Fund can be proved to hold for 0. Upon $\mathrm{Od}(c)$ we obviously have $\mathrm{Trans}(c;c)$ and $\mathrm{Alt}(c;c)$; $\mathrm{Fund}(c;c)$ follows from 1.11 and 1.5. As to 2.3 we have upon the condition $(\mathfrak{x})(\mathfrak{x} \in m \to \mathrm{Od}(\mathfrak{t}(\mathfrak{x})))$ first directly $\mathrm{Trans}(\sum_{\mathfrak{x}}(m, \mathfrak{t}(\mathfrak{x})))$ and from this and our premise follows $\mathrm{Od}(\sum_{\mathfrak{x}}(m, \mathfrak{t}(\mathfrak{x})))$ by 1.5 and 1.13.

2.3 gives in particular

2.4 $$(x)(x \in m \to \mathrm{Od}(x)) \to \mathrm{Od}(\sum m)$$

«The sum of the elements of a set of ordinals is again an ordinal».

Defining

Df 2.1 $$c' = c;c$$

we immediately have

2.5 $$a \in a', \ a' \neq 0, \ \mathrm{Od}(a) \to \mathrm{Od}(a')$$

and by 1.11

2.6 $$\mathrm{Od}(a) \ \& \ \mathrm{Od}(b) \ \& \ a' = b' \to a = b.$$

A nearer characterization of 0, c' and $\sum m$ is given by the formulas

2.7 $$\mathrm{Od}(c) \to 0 = c \lor 0 \in c,$$

«0 is the lowest ordinal».

2.8 $$\mathrm{Od}(c) \ \& \ a \in c \to a' = c \lor a' \in c.$$

2.8 together with 2.5 says that a' is for every ordinal a the next higher ordinal; we call it the *successor* of a.

2.9 $$(x)(x \in m \to \mathrm{Od}(x)) \cdot\!\to\!\cdot (x)(x \in m \to x = \sum m \lor x \in \sum m)$$

2.10 $$\mathrm{Od}(c) \ \& \ (x)(x \in m \to \mathrm{Od}(x) \ \& \ x \subseteq c) \to \sum m = c \lor \sum m \in c.$$

«For every set of ordinals m, $\sum m$ is at least as high an ordinal as any element of m, and it is the lowest one having that property».

The proofs of the formulas 2.7–2.10 go with 1.5, 1.8, 1.11 and 2.1–2.4.

From 2.8 we also directly derive

2.11 $$\mathrm{Od}(c) \;\&\; (x)(x' \neq c) \;\rightarrow\; (x)(x \in c \rightarrow x' \in c).$$

Thus defining

Df 2.2 $$\mathrm{Suc}(c) \;\leftrightarrow\; (Ex)(\mathrm{Od}(x) \;\&\; x' = c)$$
«c is a successor number».

and

Df 2.3 $$\mathrm{Lim}(c) \;\leftrightarrow\; \mathrm{Od}(c) \;\&\; c \neq 0 \;\&\; (x)(x \in c \rightarrow x' \in c)$$
«c is a limit number»

we have

2.12 $$\mathrm{Od}(c) \;\rightarrow\; c = 0 \vee \mathrm{Suc}(c) \vee \mathrm{Lim}(c)$$

and also

2.13 $$\mathrm{Lim}(c) \;\rightarrow\; c \neq 0 \;\&\; c \neq a'.$$

Note that an ordinal c is a successor if and only if it has a highest element.
The difference between successors and limit numbers appears pregnantly in connection with the sum operator. Indeed we easily get

2.14 $$\mathrm{Od}(c) \rightarrow \textstyle\sum c' = c, \quad \mathrm{Lim}(c) \rightarrow \textstyle\sum c = c.$$

As a consequence of 2.4 and 2.9 we have that for every set of ordinals m with no highest element, $\sum m$ is a higher ordinal than every element of m.

From this we can infer that there does not exist a set of all ordinals, or in other words, that $\{x \mid \mathrm{Od}(x)\}$ is not represented. For, if m were the set of all ordinals then, since to every ordinal exists a higher one, $\sum m$ would be an ordinal higher than every element of m, what obviously is contradictory. Thus here already the necessity of distinguishing classes and sets comes to appear.

We further note here, that according to 2.7, 2.8 and 2.10 we have the following properties of any ordinal c: 1) 0 is either equal to c or an element of c; 2) if a is an element of c, then a' is either equal to c or an element of c; 3) if m is a subset of c, then $\sum m$ is either equal to c or an element of c.

It can be proved on the other hand that every set having the said three properties is an ordinal. Zermelo in his mentioned unpublished theory of ordinals, which makes no use of the concept of order, defined ordinals by these properties.

Let us note that we have till now no means for proving the existence of a limit number. This will only be possible when we shall have introduced the axiom of infinity.

III, § 3. Fundaments of number theory

From the laws of ordinal theory, derived in the last sections, we can obtain by the way of specialization the fundaments of number theory. We observe that for this purpose not our full axiomatic basis is required. Indeed surveying the last two sections we see that in all occurring proofs except those of formulas containing $\sum_{\mathfrak{x}}(m, \mathfrak{t}(\mathfrak{x}))$, $\sum_{x}(m, x)$, $\sum m$, the axiom A 3 was used only in the form of the Aussonderungstheorem, so that it would be sufficient for them to have instead of A 3 the weaker axiomatical assumption of the existence of the set $\mathrm{a} \cap A$ with its characterizing property II, 2.3. The frame thus delimited, which we may call the *weakened general set theory*, allows already to derive number theory. In fact we shall use for this only those theorems of the last sections which are proved in the said frame.

Natural numbers can be defined as special ordinals, according to the following definition:

Df 3.1 $\quad \mathrm{Nu}(n) \cdot\leftrightarrow\cdot \mathrm{Od}(n) \ \& \ (n=0 \vee \mathrm{Suc}(n)) \ \& \ (x)(x \in n \rightarrow$
$\rightarrow x=0 \vee \mathrm{Suc}(x))$.

«A natural number is an ordinal such that itself and every element of it is either 0 or a successor.»

From this definition we immediately get, by the formulas 2.1, 2.5 and 2.6

3.1 $\quad \mathrm{Nu}(0),\ \mathrm{Nu}(a) \rightarrow \mathrm{Nu}(a'),\ \mathrm{Nu}(a) \ \& \ \mathrm{Nu}(b) \ \& \ a'=b' \rightarrow$
$\rightarrow a=b,\ a' \neq 0$

as also

3.2 $$\mathrm{Nu}(a) \mathbin{\&} b \in a \rightarrow \mathrm{Nu}(b)$$

and with this

3.3 $$\mathrm{Nu}(a) \rightarrow a=0 \vee (Ey)(\mathrm{Nu}(y) \mathbin{\&} y'=a).$$

With 3.1 we have already all Peano axioms up to that of (*complete*) *numeral induction:*

3.4 $$0 \in A \mathbin{\&} (x)(\mathrm{Nu}(x) \mathbin{\&} x \in A \rightarrow x' \in A) \cdot\!\rightarrow\!\cdot \mathrm{Nu}(a) \rightarrow a \in A.$$

We derive this from 1.17. Substituting here $\mathrm{Nu}(c) \rightarrow c \in A$ for $\mathfrak{A}(c)$ and using $\mathrm{Nu}(c) \rightarrow \mathrm{Od}(c)$ and 3.2, we get

$$(x)(\mathrm{Nu}(x) \mathbin{\&} (z)(z \in x \rightarrow z \in A) \rightarrow x \in A) \cdot\!\rightarrow\!\cdot (\mathrm{Nu}(c) \rightarrow c \in A).$$

Hence for deriving 3.4 it is sufficient to prove

[1] $$0 \in A \mathbin{\&} (x)(\mathrm{Nu}(x) \mathbin{\&} x \in A \rightarrow x' \in A) \cdot\!\rightarrow\!\cdot$$
$$\cdot\!\rightarrow\!\cdot \mathrm{Nu}(c) \rightarrow ((z)(z \in c \rightarrow z \in A) \rightarrow c \in A).$$

For this proof we use the formula 3.3; from this, with the predicate calculus and the equality axioms, we get the schema

3.5 $$\mathfrak{C}(0) \mathbin{\&} (y)(\mathrm{Nu}(y) \rightarrow \mathfrak{C}(y')) \cdot \rightarrow \cdot \mathrm{Nu}(c) \rightarrow \mathfrak{C}(c).$$

Applying this to the case of $\mathfrak{C}(c)$ being $(z)(z \in c \rightarrow z \in A) \rightarrow c \in A$ we first observe that we have $0 \in A \rightarrow \mathfrak{C}(0)$, and so the proof of [1] comes back to that of

$$(x)\mathrm{Nu}(x) \mathbin{\&} x \in A \rightarrow x' \in A) \cdot\!\rightarrow\!\cdot (y)(\mathrm{Nu}(y) \rightarrow \mathfrak{C}(y'))$$

and thus also to the proof of

[2] $$(x)(\mathrm{Nu}(x) \mathbin{\&} x \in A \rightarrow x' \in A) \cdot\!\rightarrow\!\cdot \mathrm{Nu}(c) \mathbin{\&} (z)(z \in c' \rightarrow$$
$$\rightarrow z \in A) \rightarrow c' \in A.$$

Now [2] results as follows: We have by the predicate calculus

$$(z)(z \in c' \rightarrow z \in A) \rightarrow (c \in c' \rightarrow c \in A)$$
$$\rightarrow c \in A \qquad \text{by 2.5,}$$

and therefore

$$\mathrm{Nu}(c) \mathbin{\&} c \in A \rightarrow c' \in A) \cdot\!\rightarrow\!\cdot \mathrm{Nu}(c) \mathbin{\&} (z)(z \in c' \rightarrow z \in A) \rightarrow c' \in A,$$

and this obviously gives [2].

From 3.4, by the Church schema, we have the immediate passage to the formula schema

3.6 $\mathfrak{A}(0)\ \&\ (\mathfrak{x})(\mathrm{Nu}(\mathfrak{x})\ \&\ \mathfrak{A}(\mathfrak{x}) \rightarrow \mathfrak{A}(\mathfrak{x}')) \cdot\!\!\rightarrow\!\cdot (\mathrm{Nu}(a) \rightarrow \mathfrak{A}(a))$.

We have derived 3.4 from 1.17 which itself was obtained from the principle of least ordinal number 1.12. From this we have drawn the formulas for the μ-operator 1.14–1.16, which gave an explicit formulation of that principle. These formulas can now also directly be applied to number theory by means of the definition

Df 3.2 $$\dot{\mu}A = \mu(A \cap \{x \mid \mathrm{Nu}(x)\}).$$

By this way we come to the formulas expressing the *principle of least natural number*

3.7 $$\mathrm{Nu}(c)\ \&\ c \in A \rightarrow \mathrm{Nu}(\dot{\mu}A)\ \&\ \dot{\mu}A \in A$$

3.8 $$\mathrm{Nu}(c)\ \&\ c \in A \rightarrow \dot{\mu}A = c \vee \dot{\mu}A \in c$$

3.9 $$(x)\mathrm{Nu}(x) \rightarrow x \notin A) \rightarrow \dot{\mu}A = 0.$$

By the Church schema we obviously can pass from these formulas to the corresponding formula schemata for a predicate $\mathfrak{A}$ instead of the class variable A [1]).

The main thing still required for developing number theory is the method of introducing functions by primitive recursion. As is well known this requirement, which was first explicitly noticed by Dedekind, in most formalizations of number theory gives rise to some complications, if not the recursive definition is taken as a primitive rule. Here we can use the Dedekind method with the simplification that the relation $\leqslant$ between numbers is directly expressible by $a \in b \vee a = b$. At the same time we make use of the fact that primitive recursion can be reduced – whenever ordered pairs are available as individuals – to iteration [2]).

---

[1]) The operator $\mu_x\mathfrak{A}(x)$ in [Hilbert and Bernays 1934/39]—in distinction from our present more general μ-symbol for ordinals—is here to be defined by the expression $\dot{\mu}\{x \mid \mathfrak{A}(x)\}$.

[2]) This was first observed, as it seems, by S. C. Kleene [1933] and has been employed in several newer publications, in particular of Quine's school, for the reduction of primitive recursion to the ancestral.

In fact we shall derive the theorem justifying primitive recursion from an iteration theorem, which we now come to discuss.

III, § 4. Iteration. Primitive recursion

For formulating the iteration theorem we need the concept of a *sequence* which in the following will occur in many contexts. By a sequence we understand a functional set whose domain is an ordinal:

Df 4.1 $$\mathrm{Sq}(s) \leftrightarrow \mathrm{Ft}(s) \;\&\; \mathrm{Od}(\Delta_1 s).$$

The elements of the converse domain $\Delta_2 s$ of a sequence s are called the *members* of s. (Thus we have to distinguish between the members and the elements of a sequence.) Note that the empty set by Df 4.1 is also a sequence.

We now further define the concept of an *iteration sequence for F, starting from a*:

Df 4.2 $$\mathrm{It}(s, a, F) \leftrightarrow \mathrm{Sq}(s) \;\&\; \mathrm{Nu}(\Delta_1 s) \;\&\; s^{\text{\textquoteleft}}0 = a \;\&\; \\ \&\; (x)(x' \in \Delta_1 s \rightarrow \langle s^{\text{\textquoteleft}}x, s^{\text{\textquoteleft}}x'\rangle \in F)$$

and by means of this the *numeral iterator of F on a*:

Df 4.3 $$\mathsf{J}(F, a) \equiv \{xy \mid (Ez)(\mathrm{It}(z, a, F) \;\&\; \langle x, y\rangle \in z)\}.$$

Intuitively the iterator of F on a can be characterized as the class of pairs $\langle n, b\rangle$ with n being a natural number and b being the final member a_n of some sequence $a_0, a_1, a_2, \ldots, a_n$, where $a_0 = a$ and $\langle a_{i-1}, a_i\rangle$ belongs to F for $i = 1, 2, \ldots, n$.

Now the iteration theorem says that, if A is a class of which a is an element and F is a function mapping A into A, then the iterator $\mathsf{J}(F, a)$ is a function H with the domain $\{x \mid \mathrm{Nu}(x)\}$ which satisfies the equations

$$H^{\text{\textquoteleft}}0 = a, \quad \mathrm{Nu}(n) \rightarrow H^{\text{\textquoteleft}}n' = F^{\text{\textquoteleft}}(H^{\text{\textquoteleft}}n)$$

and whose values belong to A; formally

4.1 $$a \in A \;\&\; \mathrm{Ft}(F) \;\&\; \Delta_1 F \equiv A \;\&\; \Delta_2 F \subseteq A \;\&\; H \equiv \mathsf{J}(F, a) \cdot\!\rightarrow\!\cdot \\ \cdot\!\rightarrow\!\cdot \mathrm{Ft}(H) \;\&\; \Delta_1 H \equiv \{x \mid \mathrm{Nu}(x)\} \;\&\; \Delta_2 H \subseteq A \;\&\; H^{\text{\textquoteleft}}0 = a \;\&\; \\ \&\; (x)(\mathrm{Nu}(x) \rightarrow H^{\text{\textquoteleft}}x' = F^{\text{\textquoteleft}}(H^{\text{\textquoteleft}}x)).$$

For the indication of the proof let us denote briefly the expression $a \in A$ & $\mathrm{Ft}(F)$ & $\Delta_1 F \equiv A$ & $\Delta_2 F \subseteq A$ by $\mathfrak{C}(F, a, A)$. We then are to prove

[1] $\mathfrak{C}(F, a, A)$ & $\mathrm{Nu}(n) \cdot\!\!\rightarrow\!\!\cdot (Ez)(\mathrm{It}(z, a, F)$ & $\Delta_1 z = n'$ & $z^{\epsilon}n \in A)$,

[2] $\mathfrak{C}(F, a, A)$ & $\mathrm{It}(s, a, F)$ & $\mathrm{It}(t, a, F)$ & $\mathrm{Nu}(n) \cdot\!\!\rightarrow\!\!\cdot$
$\cdot\!\!\rightarrow\!\!\cdot (x)(y)(\langle n, x\rangle \in s$ & $\langle n, y\rangle \in t \rightarrow x = y)$.

For both formulas the derivation goes by numeral induction with respect to n; to the first one we use the formula

$$\mathrm{It}(s, a, F) \;\&\; \Delta_1 s = n' \;\&\; \langle s^{\epsilon}n, b\rangle \in F \cdot\!\!\rightarrow\!\!\cdot \mathrm{It}((s; \langle n', b\rangle), a, F).$$

With the help of [1] and [2], using Df 4.2 and Df 4.3, we get the formula 4.1 to be proved.

Concerning the iterator we note the following formula sometimes to be applied

4.2 $$\mathrm{Crs}(F) \;\&\; a \notin \Delta_2 F \rightarrow \mathrm{Crs}(\mathrm{J}(F, a)).$$

The proof goes by deriving with numeral induction

$$\mathrm{Crs}(F) \;\&\; a \notin \Delta_2 F \cdot\!\!\rightarrow\!\!\cdot \mathrm{Nu}(c) \;\&\; b \in c \rightarrow (\overline{Ex})\,(\langle b, x\rangle \in \mathrm{J}\,(F, a) \;\&$$
$$\& \;\langle c, x\rangle \in \mathrm{J}\,(F, a)).$$

By application of the iterator we are able to define the elementary arithmetic functions: Indeed defining

Df 4.4 $$m + n = \mathrm{J}(\{xy \mid y = x'\}, m)^{\epsilon}n,$$

Df 4.5 $$m \cdot n = \mathrm{J}(\{xy \mid y = x + m\}, 0)^{\epsilon}n,$$

Df 4.6 $$m^n \;= \mathrm{J}(\{xy \mid y = x \cdot m\}, 0')^{\epsilon}n,$$

we obtain by 4.1, with substituting $\{x \mid \mathrm{Nu}(x)\}$ for A, the formulas constituting the ordinary recursive definitions of these functions:

4.3 $$\mathrm{Nu}(m) \;\&\; \mathrm{Nu}(n) \rightarrow \mathrm{Nu}(m+n) \;\&\; \mathrm{Nu}(m \cdot n) \;\&\; \mathrm{Nu}(m^n)$$

4.4 $$\mathrm{Nu}(m) \;\&\; \mathrm{Nu}(n) \rightarrow m + 0 = m \;\&\; m + n' = (m+n)'$$

4.5 $$\mathrm{Nu}(m) \;\&\; \mathrm{Nu}(n) \rightarrow m \cdot 0 = 0 \;\&\; m \cdot n' = (m \cdot n) + m$$

4.6 $$\mathrm{Nu}(m) \;\&\; \mathrm{Nu}(n) \rightarrow m^0 = 0' \;\&\; m^{n'} = m^n \cdot m.$$

From the iteration theorem we now pass to the *schema of primitive recursion.* Let us consider the schema for the case of one parameter; the corresponding schema with more parameters can immediately be reduced to the former by comprehending parameters into one $\mathfrak{k}$-tuplet. We need not restrict the schema to the definition of functions with number values but can state it in the following more general form:

4.7 Upon the condition that we have for some formulas $\mathfrak{A}(c)$, $\mathfrak{C}(c)$ and some terms $\mathfrak{t}(c)$, $\mathfrak{p}(a, b, c)$

$$\text{(C)} \quad \begin{aligned} &\mathfrak{A}(a) \to \mathfrak{C}(\mathfrak{t}(a)) \\ &\mathfrak{A}(a) \,\&\, \mathfrak{C}(b) \,\&\, \mathrm{Nu}(m) \to \mathfrak{C}(\mathfrak{p}(m, b, a)), \end{aligned}$$

a function symbol $\mathfrak{f}(a, m)$ can be introduced, with the formulas

[1] $\mathfrak{A}(a) \to \mathfrak{f}(a, 0) = \mathfrak{t}(a)$

[2] $\mathfrak{A}(a) \,\&\, \mathrm{Nu}(m) \to \mathfrak{C}(\mathfrak{f}(a, m)) \,\&\, \mathfrak{f}(a, m') = \mathfrak{p}(m, \mathfrak{f}(a, m), a)$

taken as its recursive definition.

For justifying this procedure we show that we can set up an explicit definition for $\mathfrak{f}(a, m)$ in such a way that [1] and [2] are provable under the condition (C). To this end we apply the iteration theorem 4.1 with substituting

for a : $\langle 0, \mathfrak{t}(a)\rangle$

for A: $\{\mathfrak{x}\mathfrak{y} \mid \mathrm{Nu}(\mathfrak{x}) \,\&\, \mathfrak{C}(\mathfrak{y})\}$

for F: $\{\mathfrak{u}\mathfrak{v} \mid (E\mathfrak{z})(E\mathfrak{w})(\mathrm{Nu}(\mathfrak{z}) \,\&\, \mathfrak{C}(\mathfrak{w}) \,\&$
$\&\, \mathfrak{u} = \langle \mathfrak{z}, \mathfrak{w}\rangle \,\&\, \mathfrak{v} = \langle \mathfrak{z}', \mathfrak{p}(\mathfrak{z}, \mathfrak{w}, a)\rangle)\}$.

Denoting the two class terms by $\mathfrak{K}$ and $\mathfrak{F}(a)$ we have by (C)

$$\mathfrak{A}(a) \to \langle 0, \mathfrak{t}(a)\rangle \in \mathfrak{K} \,\&\, \mathrm{Ft}(\mathfrak{F}(a)) \,\&\, \Delta_1\mathfrak{F}(a) \equiv \mathfrak{K} \,\&\, \Delta_2\mathfrak{F}(a) \subseteq \mathfrak{K}.$$

Thus by 4.1 we obtain

$$\begin{aligned} \mathfrak{A}(a) \,\&\, H \equiv \mathrm{J}(\mathfrak{F}(a), \langle 0, \mathfrak{t}(a)\rangle) \to \mathrm{Ft}(H) \,\&\, \Delta_1 H \equiv \{\mathfrak{x} \mid \mathrm{Nu}(\mathfrak{x})\} \,\& \\ \&\, \Delta_2 H \subseteq \mathfrak{K} \,\&\, H`0 = \langle 0, \mathfrak{t}(a)\rangle \,\&\, (\mathfrak{x})(\mathrm{Nu}(\mathfrak{x}) \to \\ \to H`\mathfrak{x}' = \mathfrak{F}(a)`(H`\mathfrak{x})). \end{aligned}$$

The elements of the iterator $\mathrm{J}(\mathfrak{F}(a), \langle 0, \mathfrak{t}(a)\rangle)$ are triplets of the form $\langle \mathfrak{n}, \langle \mathfrak{n}, \mathfrak{k} \rangle\rangle$, with a natural number $\mathfrak{n}$. Indeed by numeral induction we can prove

$$\mathrm{Nu}(m) \rightarrow (E\mathfrak{z})(\mathsf{J}(\mathfrak{F}(a), \langle 0, \mathfrak{t}(a)\rangle)^{\iota} m = \langle m, \mathfrak{z}\rangle).$$

Now defining

[3] $$\mathfrak{f}(a, m) = \iota_{\mathfrak{x}}(\langle m, \mathfrak{x}\rangle = \mathsf{J}(\mathfrak{F}(a), \langle 0, \mathfrak{t}(a)\rangle)^{\iota} m)$$

we get

$$\mathfrak{A}(a) \rightarrow \langle 0, \mathfrak{f}(a, 0)\rangle = \langle 0, \mathfrak{t}(a)\rangle,$$

$$\begin{aligned} \mathfrak{A}(a) \,\&\, \mathrm{Nu}(m) &\rightarrow \langle m', \mathfrak{f}(a, m')\rangle = \mathsf{J}(\mathfrak{F}(a), \langle 0, \mathfrak{t}(a)\rangle)^{\iota} m' \\ &\rightarrow \langle m', \mathfrak{f}(a, m')\rangle = \mathfrak{F}(a)^{\iota}(\mathsf{J}(\mathfrak{F}(a), \langle 0, \mathfrak{t}(a)\rangle)^{\iota} m \\ &\rightarrow \langle m', \mathfrak{f}(a, m')\rangle = \mathfrak{F}(a)^{\iota} \langle m, \mathfrak{f}(a, m)\rangle \\ &\rightarrow \langle m', \mathfrak{f}(a, m')\rangle = \langle m', \mathfrak{p}(m, \mathfrak{f}(a, m), a)\rangle, \end{aligned}$$

what by II, 1.13 gives [1] and [2]. –

The concept of an iteration sequence can in particular be used to define the *transitive closure* of a class

Df 4.7 $\lceil A \rceil \equiv \{z \mid (Eu)(Ev)(\mathrm{It}(u, v, \{xy \mid y \in x\}) \,\&\, v \in A \,\&\, z \in \Delta_2 u)\}$

«z belongs to $\lceil A \rceil$ if it is a member of a sequence $a_0, a_1, a_2, \ldots, a_n$, where $a_0 \in A$ and $a_k \in a_{k-1}$ $(k = 1, 2, \ldots, n)$».

By the transitive closure of a set a we understand that one of a^*, i.e. $\lceil a^* \rceil$.

For formulating the characteristic properties of the transitive closure, we define Trans(A) by extending our former Df 1.1 to classes. Now using Df 4.2 we can prove

4.8 $$A \subseteq \lceil A \rceil, \quad \mathrm{Trans}(\lceil A \rceil)$$

4.9 $$\mathrm{Trans}(C) \,\&\, A \subseteq C \rightarrow \lceil A \rceil \subseteq C.$$

«$\lceil A \rceil$ is transitive, and A is a subclass of $\lceil A \rceil$ as also of every transitive class of which A is a subclass, i.e. briefly $\lceil A \rceil$ is the smallest transitive class containing A as a subclass.»

The proof of 4.9 goes by deriving

$$\begin{aligned} \mathrm{Trans}(C) \,\&\, c \in C \,\&\, \mathrm{Nu}(n) \rightarrow (z)((Eu)\,(\mathrm{It}(u, c, \{xy \mid y \in x\}) \,\&\, \\ \&\, \langle n, z\rangle \in u) \rightarrow z \in C) \end{aligned}$$

with numeral induction.

Our way of defining the transitive closure is adapted to the frame of weakened general set theory. With employing the sum operator another method of introducing the transitive closure can be used, by which we have first to define the transitive closure of a set.

This possibility is due to the circumstance that $\overline{a^*|}$ is the union of the sets a, $\sum a$, $\sum(\sum a)$, ... which are the values of the iterator of the function $\{xy \mid y = \sum x\}$ on a, so that we have

4.10 $$\overline{a^*|} \equiv \bigcup \Delta_2 \mathsf{J}(\{xy \mid y = \textstyle\sum x\}, a).$$

For the proof of this class equality it is in virtue of 4.8 and 4.9 sufficient to show that the class $\bigcup \Delta_2 \mathsf{J}(\{xy \mid y = \sum x\}, a)$ – let us write instead briefly $\mathfrak{K}(a)$ – has the characterizing properties of the transitive closure:

[1] $$a^* \subseteq \mathfrak{K}(a), \quad \mathrm{Trans}(\mathfrak{K}(a))$$

[2] $$\mathrm{Trans}(B) \mathbin{\&} a^* \subseteq B \rightarrow \mathfrak{K}(a) \subseteq B.$$

For deriving these formulas we have of course to use 4.1. Substituting here V for A and $\{xy \mid y = \sum x\}$ for F and the class-term $\mathsf{J}(\{xy \mid y = \sum x\}, a)$ – which we briefly denote by $\mathfrak{B}(a)$ – for H, we get

[3] $$\mathrm{Ft}(\mathfrak{B}(a)) \mathbin{\&} \Delta_1 \mathfrak{B}(a) \equiv \{x \mid \mathrm{Nu}(x)\} \mathbin{\&} \mathfrak{B}(a)\text{‘}0 = a \mathbin{\&}$$
$$\mathbin{\&} (z)(\mathrm{Nu}(z) \rightarrow \mathfrak{B}(a)\text{‘}z' = \textstyle\sum \mathfrak{B}(a)\text{‘}z).$$

Further by Df I, 3.8 we have

[4] $$\bigcup \Delta_2 \mathfrak{B}(a) \equiv \{u \mid (Ev)(\mathrm{Nu}(v) \mathbin{\&} u \in \mathfrak{B}(a)\text{‘}v)\}.$$

From [3] and [4] we first get the formulas [1], using Df II, 1.5, and [2] is obtained by deriving

$$\mathrm{Trans}(B) \mathbin{\&} a^* \subseteq B \cdot\!\rightarrow\!\cdot \mathrm{Nu}(n) \rightarrow (\mathfrak{B}(a)\text{‘}n)^* \subseteq B$$

with numeral induction on n.

Moreover the transitive closure of an arbitrary class is expressible by the closure of a represented class, since we have

4.11 $$\overline{A|} \equiv \{y \mid (Ex)(x \in A \mathbin{\&} (y = x \vee y \in \overline{x^*|}))\}.$$

In fact the characterizing properties of $\overline{A|}$ can be shown to hold for the class

$$\{y \mid (Ex)(x \in A \ \& \ (y = x \vee y \in \overline{x^*|}))\},$$

using that they hold for any class $\overline{c^*|}$.

III, § 5. Finite sets and classes

The concept of finiteness in set theory is subject to certain complications, because there are many different ways of defining it, all intuitively motived, which however can be shown to be equivalent only with the help of the axiom of choice. Interesting results in this respect have been attained by A. Tarski, A. Mostowski and A. Lindenbaum [1925a, 1938, 1938, 1945].

In particular we have the possibility of either defining finity or else infinity in a positive sense. Infinity is for instance positively defined by Dedekind's well known definition according to which a set is infinite if there exists a one-to-one correspondence of it with a proper subset. We shall later on come back to these questions.

Here in the theory of finite sets we define the concept of finiteness in a positive and relatively elementary way:

Df 5.1 $\quad \mathrm{Fin}(a) \leftrightarrow (Ex)(\mathrm{Nu}(x) \ \& \ x \sim a).$

«A set is finite if there exists a one-to-one correspondence between it and a natural number».

Df 5.2 $\quad \mathrm{Fin}(A) \leftrightarrow (Ex)(\mathrm{Fin}(x) \ \& \ x^* \equiv A).$

«A class is finite if it is represented by a finite set».

That no unnecessary restriction is included in the last definition appears from the provability of the formula

5.1 $\quad \mathrm{Nu}(n) \ \& \ n^* \mathrel{\overline{\underset{C}{\ }}} A \rightarrow \mathrm{Fin}(A)$

which consists in virtue of II, 4.5. (Cf. the remark on p. 99.)

In order to derive the familiar properties of finite sets and in particular those connected with the concept of the number of elements – let us briefly say the "*multitude*" – of a finite set, we need two preliminary theorems. The first one is

5.2 $\quad \mathrm{Nu}(n) \ \& \ s \subset n \rightarrow (Ex)(\mathrm{Nu}(x) \ \& \ x \in n \ \& \ x \sim s).$

The proof goes by applying numeral induction to the predicate of n:

$$(u)(u \subset n \rightarrow (Ex)(\mathrm{Nu}(x) \;\&\; x \in n \;\&\; x \sim u))$$

and using the formulas

$$\mathrm{Nu}(n) \;\&\; a \in n \rightarrow 0 \in n,$$
$$t \subset n' \rightarrow t = n \vee t \subset n \vee (t = (t \cap n) \cup [n] \;\&\; (t \cap n) \subset n),$$
$$\mathrm{Nu}(m) \;\&\; \mathrm{Nu}(n) \rightarrow (m \in n \rightarrow m' \in n').$$

The other preliminary theorem is

5.3 $$\mathrm{Nu}(k) \;\&\; \mathrm{Od}(n) \;\&\; k \sim n \rightarrow k = n.$$

This follows by applying the principle of least natural number to the class

$$\{x \,|\, \mathrm{Nu}(x) \;\&\; (Ey)(\mathrm{Od}(y) \;\&\; x \sim y \;\&\; x \neq y)\},$$

using 5.2 and 1.10.

As consequences of 5.2 and 5.3 we have

5.4 $$\begin{cases} a \subseteq b \;\&\; \mathrm{Fin}(b) \rightarrow \mathrm{Fin}(a) \\ A \subseteq B \;\&\; \mathrm{Fin}(B) \rightarrow \mathrm{Fin}(A) \end{cases}$$

5.5 $$\mathrm{Od}(n) \cdot\rightarrow\cdot \mathrm{Fin}(n) \leftrightarrow \mathrm{Nu}(n)$$

5.6 $$a \subset b \;\&\; (\mathrm{Fin}(a) \vee \mathrm{Fin}(b)) \cdot\rightarrow\cdot \overline{a \sim b}.$$

Thus there is no one-to-one correspondence between a finite set and a proper subset of it. Defining

Df 5.3 $$\mathrm{mlt}(a) = \iota_x(\mathrm{Nu}(x) \;\&\; x \sim a)$$
«the multitude of a»

we have

5.7 $$\mathrm{Fin}(a) \cdot\rightarrow\cdot (x)(\mathrm{Nu}(x) \;\&\; x \sim a \leftrightarrow x = \mathrm{mlt}(a)),$$

what of course entails

5.8 $$\mathrm{Fin}(a) \cdot\rightarrow\cdot \mathrm{Nu}(\mathrm{mlt}(a)) \;\&\; a \sim \mathrm{mlt}(a)).$$

Further we have the following theorems of the theory of multitudes which we state here without indication of the proofs, that go by formalizing the familiar intuitive arguments:

5.9 $$a \subset b \;\&\; \mathrm{Fin}(b) \cdot\rightarrow\cdot \mathrm{mlt}(a) \in \mathrm{mlt}(b)$$

5.10 $$\mathrm{Fin}(a) \;\&\; \mathrm{Fin}(b) \cdot\rightarrow\cdot \mathrm{Fin}(a \cup b) \;\&\; (a \cap b = 0 \rightarrow \mathrm{mlt}(a \cup b) = \\ = \mathrm{mlt}(a) + \mathrm{mlt}(b)),$$

5.11 $\mathrm{Fin}(a) \,\&\, \mathrm{Fin}(b) \rightarrow \mathrm{Fin}(a \times b) \,\&\, \mathrm{mlt}(a \times b) = \mathrm{mlt}(a) \cdot \mathrm{mlt}(b)$,

5.12 $\mathrm{Fin}(a) \,\&\, \mathrm{Fin}(b) \rightarrow \mathrm{Fin}(a^{*\underline{b}}) \,\&\, (Ex)(x^* \equiv a^{*\underline{b}} \,\&$
$\& \,\mathrm{mlt}(x) = \mathrm{mlt}(a)^{\mathrm{mlt}(b)})$

where $\mathrm{mlt}(a)^{\mathrm{mlt}(b)}$ is the value of the arithmetical function m^n defined by Df 4.6.

We also obtain

5.13 $\mathrm{Fin}(a) \,\&\, (x)(x \in a \rightarrow \mathrm{Fin}(x)) \rightarrow \mathrm{Fin}(\sum a)$.

«The union of finitely many finite sets is again finite».

The derivation is by numeral induction with respect to the multitude of a, i.e. by deriving the formula (obviously equivalent with 5.13)

$$\mathrm{Nu}(n) \cdot\!\rightarrow\!\cdot (z)(z \sim n) \,\&\, (x)(x \in z \rightarrow \mathrm{Fin}(x)) \rightarrow \mathrm{Fin}(\textstyle\sum z))$$

by numeral induction with respect to n, using 5.10.

Thus the usual theory of multitudes can here be obtained, without employing the Anzahldefinition of Frege.

Finally we note the theorem that in every (non empty) finite set of ordinals there is a highest ordinal:

5.14 $c \neq 0 \,\&\, \mathrm{Fin}(c) \,\&\, (x)(x \in c \rightarrow \mathrm{Od}(x)) \cdot\!\rightarrow\!\cdot (Ey)(y \in c \,\&$
$\& \,(x)(x \in c \rightarrow x \subseteq y))$.

The proof goes by means of numeral induction with respect to the multitude of c.

Remark: The proof of 5.1, which goes quickly with using II, 3.8, can be given also in the frame of weakened general set theory, as follows: by numeral induction with respect to n we get

$$\mathrm{Ft}(F) \,\&\, \mathrm{Nu}(n) \,\&\, n^* \subseteq \Delta_1 F \rightarrow \mathrm{Rp}(\{y \,\|\, (Ex)(x \in n \,\&\, \langle x, y \rangle \in F)\}).$$

By application of this formula we obtain successively

$$\begin{aligned} \mathrm{Nu}(n) \,\&\, n^* \,\overline{\urcorner}_{C}\, A \rightarrow A \equiv \{y \,\|\, (Ex)(x \in n \,\&\, \langle x, y \rangle \in C)\} \,\&\, \mathrm{Rp}(A) & \\ \rightarrow (Ez)(n^* \,\overline{\urcorner}_{C}\, z^* \,\&\, z^* \equiv A) & \\ \rightarrow (Ez)(Eu)(n^* \,\overline{\urcorner}_{u^*}\, z^* \,\&\, z^* \equiv A), & \quad \text{by II, 4.5} \\ \rightarrow (Ez)(n \sim z \,\&\, z^* \equiv A) & \\ \rightarrow \mathrm{Fin}(A), & \quad \text{by Df 5.2 and Df 5.1.} \end{aligned}$$

CHAPTER IV

TRANSFINITE RECURSION

IV, § 1. The general recursion theorem

The method of recursive definition is not restricted to number theory, but can be established as a general procedure of ordinal theory. In this extended form, we call it *transfinite recursion.* This again can be based on a still more general theorem which we call the *general recursion theorem*. This theorem was first stated and proved exactly by von Neumann [1928a], who at once stressed that besides the original Zermelo axioms also the assertion of the axiom of replacement has here to be used. Owing to this circumstance we have to deal now, in considering general recursion and its application, again with our full axiom system A 1–A 3 of general set theory.

For the statement of the general recursion theorem some concepts have to be defined. We first extend our concept of sequence to classes, calling a function S whose domain is either an ordinal or $\{x \mid \mathrm{Od}(x)\}$ a *sequential class*:

Df 1.1 $\mathrm{Sq}(S) \leftrightarrow \mathrm{Ft}(S) \,\&\, ((Ex)(\mathrm{Od}(x) \,\&\, \Delta_1 S \equiv x^*) \vee \Delta_1 S \equiv \{x \mid \mathrm{Od}(x)\})$.

According to this definition indeed every sequence as defined by III, Df 4.1 represents a sequential class, so that we have

1.1 $$\mathrm{Sq}(s) \leftrightarrow \mathrm{Sq}(s^*).$$

A sequence t which is a subset of a sequential class S will be called a *subsequence* of S. If S is a sequential class and n an element of its domain, then there is a unique subsequence of S whose domain is n; we call it the *n-segment* of S. The formal definition can be given, using II, Df 3.1, by

Df 1.2 $$\mathrm{sg}(S, n) = \underset{x}{\Gamma}(n, \langle x, S^{\prime}x \rangle).$$

The n-segment of a sequence is defined by

Df 1.3 $$\mathrm{sg}(s, n) = \mathrm{sg}(s^*, n).$$

Note that by Df 1.2 $\mathrm{sg}(S, n)$ is also defined for the case $\Delta_1 S \subseteq n^*$; namely for $k \in n$, $k \notin \Delta_1 S$ we have $\mathrm{sg}(S, n)`k = 0$. The same applies to Df 1.3.

Concerning these concepts we note for later use the formulas

1.2 $\quad \mathrm{Sq}(S) \mathbin{\&} \mathrm{Od}(k) \mathbin{\&} n \in k \rightarrow \mathrm{sg}(S, n) = \mathrm{sg}(\mathrm{sg}(S, k), n)$

«When n is a lower ordinal than the ordinal k, then the n-segment of a sequential class S is also the n-segment of the k-segment of S»,

1.3 $\quad \mathrm{Od}(c) \mathbin{\&} \mathrm{Sq}(s) \mathbin{\&} \mathrm{Sq}(t) \mathbin{\&} (x)(x \in c \rightarrow s`x = t`x) \rightarrow$
$\rightarrow \mathrm{sg}(s, c) = \mathrm{sg}(t, c)$

which result by Df 1.3, II, Df 3.1.

We shall call a sequence whose members belong to a class C briefly a *C-sequence*. Further a function F will be said to *progress in C* if it assigns to every C-sequence an element of C. The formal definition is:

Df 1.4 $\quad \mathrm{Prog}(F, C) \leftrightarrow \mathrm{Ft}(F) \mathbin{\&} \Delta_1 F \equiv \{x \mid \mathrm{Sq}(x) \mathbin{\&} \Delta_2 x^* \subseteq C\} \mathbin{\&}$
$\mathbin{\&} \Delta_2 F \subseteq C.$

Note that $\mathrm{Prog}(F, C)$ entails that $F \not\equiv \Lambda \mathbin{\&} C \not\equiv \Lambda$ since the null set, being at all events a C-sequence, is by $\mathrm{Prog}(F, C)$ in the domain of F.

Moreover a sequential class S or a sequence s will be called *adapted to F* if, for each element n of its domain, $S`n$ or $s`n$ is assigned by F to the n-segment of S or s:

Df 1.5 $\quad \mathrm{Adp}(S, F) \leftrightarrow \mathrm{Sq}(S) \mathbin{\&} (u)(u \in \Delta_1 S \cdot\!\rightarrow\!\cdot \mathrm{sg}(S, u) \in \Delta_1 F \mathbin{\&}$
$\mathbin{\&} S`u = F`\mathrm{sg}(S, u)),$

Df 1.6 $\quad \mathrm{Adp}(s, F) \leftrightarrow \mathrm{Adp}(s^*, F).$

For this concept we have the following unicity formula

1.4 $\quad \mathrm{Adp}(s, F) \mathbin{\&} \mathrm{Adp}(t, F) \mathbin{\&} c \in \Delta_1 s \cap \Delta_1 t \rightarrow s`c = t`c.$

The proof goes by means of transfinite induction III, 1.17, using the formula 1.3. We also have

1.4a $\quad \mathrm{Adp}(S, F) \mathbin{\&} c \in \Delta_1 S \rightarrow \mathrm{Adp}(\mathrm{sg}(S, c), F).$

In connection with the concept Adp(s, F) we define

Df 1.7 $\quad \mathrm{Adp}(a, b, F) \leftrightarrow (Ez)(\mathrm{Adp}(z, F) \;\&\; \langle a, b\rangle \in z)$

«b is the value assigned to a by some sequence adapted to F».

By 1.4 we immediately have, using also $\mathrm{Adp}(a, b, F) \rightarrow \mathrm{Od}(a)$

1.5 $$\mathrm{Adp}(a, b, F) \;\&\; \mathrm{Adp}(a, c, F) \rightarrow b = c.$$

Finally the class of those ordered pairs, which are in some sequence adapted to F will be called the *adaptor of* F, formally

Df 1.8 $$\mathsf{A}F \equiv \{xy \mid \mathrm{Adp}(x, y, F)\}.$$

Now the general recursion theorem can be stated by the formula

1.6a $$\mathrm{Prog}(F, C) \rightarrow \Delta_1 \mathsf{A}F \equiv \{x \mid \mathrm{Od}(x)\} \;\&\; \Delta_2 \mathsf{A}F \subseteq C \;\&\; \&\; \mathrm{Adp}(\mathsf{A}F, F),$$

or in an equivalent more handy form

1.6b $$\mathrm{Prog}(F, C) \rightarrow \mathrm{Sq}(\mathsf{A}F) \;\&\; \Delta_1 \mathsf{A}F \equiv \{x \mid \mathrm{Od}(x)\} \;\&\; \Delta_2 \mathsf{A}F \subseteq C \;\&\; \&\; (x)(\mathrm{Od}(x) \rightarrow (\mathsf{A}F)`x = F`\mathrm{sg}(\mathsf{A}F, x)).$$

The interpretation of 1.6b gives the *general recursion theorem*: For any function F which assigns to every sequence of elements of a class C again an element of C, we can define a function G, whose domain is $\{x \mid \mathrm{Od}(x)\}$ and whose value for an ordinal k is that element of C, which is assigned by F to the k-segment of G. Our proof of this theorem consists in showing that the adaptor of F is such a function G.

We first observe that by Df 1.5 and Df 1.4

[1] $$\mathrm{Prog}(F, C) \cdot\!\rightarrow\!\cdot \mathrm{Adp}(S, F) \rightarrow \Delta_2 S \subseteq C$$

and by Df 1.7 and Df 1.6

[2] $$\mathrm{Prog}(F, C) \rightarrow \Delta_2 \mathsf{A}F \subseteq C.$$

Next we use the following auxiliary formula

1.7 $$\mathrm{Sq}(s) \;\&\; (x)(x \in \Delta_1 s \rightarrow (Ez)(\mathrm{Adp}(z, F) \;\&\; x \in \Delta_1 z \;\&\; \&\; s`x = z`x)) \cdot\!\rightarrow\!\cdot \mathrm{Adp}(s, F).$$

For the proof of 1.7 let us denote by $\mathfrak{A}(s, t, b)$ the expression

$\mathrm{Adp}(t, F)$ & $b \in \Delta_1 t$ & $s^{\iota}b = t^{\iota}b$, then we have upon our premise $\mathrm{Sq}(s)$ & $(x)(x \in \Delta_1 s \to (Ez)\mathfrak{A}(s, z, x)$, and in virtue of the formula $\mathrm{Sq}(s)$ & $b \in \Delta_1 s$ & $k \in b \to k \in \Delta_1 s$:

$$(x)(y)(x \in \Delta_1 s \;\&\; y \in x \;\to\; (Eu)\mathfrak{A}(s, u, y))$$

and by the predicate calculus

$$(x)(x \in \Delta_1 s \;\to\; (Ez)\mathfrak{A}(s, z, x) \;\&\; (y)(y \in x \to (Eu)\mathfrak{A}(s, u, y))).$$

Now using 1.4 and the transitivity of equality (with respect to $s^{\iota}y$, $u^{\iota}y$, $z^{\iota}y$) we get

$$(x)(x \in \Delta_1 s \;\to\; (Ez)(\mathfrak{A}(s, z, x) \;\&\; (y)(y \in x \to s^{\iota}y = z^{\iota}y)))$$

and thus by 1.3

$$(x)(x \in \Delta_1 s \;\to\; (Ez)(\mathfrak{A}(s, z, x) \;\&\; \mathrm{sg}(s, x) = \mathrm{sg}(z, x))),$$

hence by the defining expression for $\mathfrak{A}(s, t, b)$ and Df 1.6, Df 1.5

$$(x)(x \in \Delta_1 s \;\to\; (Ez)(\mathrm{sg}(z, x) \in \Delta_1 F \;\&\; z^{\iota}x = F^{\iota}\mathrm{sg}(z, x) \;\&\; s^{\iota}x = z^{\iota}x \;\& \\ \&\; \mathrm{sg}(s, x) = \mathrm{sg}(z, x))),$$

and from this by the premise $\mathrm{Sq}(s)$ and the equality axioms $\mathrm{Adp}(s, F)$ results.

Using Df 1.7, the formula 1.7 can be transformed into

1.8 $$\mathrm{Sq}(s) \;\&\; (x)(x \in \Delta_1 s \to \mathrm{Adp}(x, s^{\iota}x, F)) \;\to\; \mathrm{Adp}(s, F).$$

A further step is to derive

1.9 $$\mathrm{Prog}(F, C) \cdot\!\to\!\cdot\, \mathrm{Od}(c) \to (Ev)\mathrm{Adp}(c, v, F).$$

The proof is by transfinite induction with respect to c: Upon the premise $\mathrm{Prog}(F, C)$ and $\mathrm{Od}(c)$ & $b \in c \to (Ev)\mathrm{Adp}(b, v, F)$ let s be the set representing the class $\{xy \mid x \in c \;\&\; \mathrm{Adp}(x, y, F)\}$. Then we have first by 1.5 and $\mathrm{Od}(c)$: $\mathrm{Sq}(s)$ and $\Delta_1 s = c$ and therefore, by 1.8, $\mathrm{Adp}(s, F)$ and thus also by $\mathrm{Prog}(F, C)$ and [1]: $\Delta_2 s \subseteq C$, hence, by Df 1.4, $s \in \Delta_1 F$. Now let t be $s;\langle c, F^{\iota}s\rangle$, then directly results

$$\mathrm{Sq}(t),\;\; \Delta_1 t = c',\;\; \mathrm{Adp}(t, F)$$

and hence

$$(Ev)\mathrm{Adp}(c, v, F).$$

The formula 1.9 gives

[3] $$\mathrm{Prog}(F, C) \to \Delta_1 \mathrm{A}F \equiv \{x \mid \mathrm{Od}(x)\}.$$

In virtue of [2] and [3] we only still have to prove, for getting 1.6, the formula

$$\mathrm{Prog}(F, C) \rightarrow \mathrm{Adp}(\mathsf{A}F, F).$$

By 1.5 and [3] we have $\mathrm{Sq}(\mathsf{A}F)$. From this and [2] we draw

$$\mathrm{Prog}(F, C) \cdot\rightarrow\cdot \mathrm{Od}(c) \rightarrow \mathrm{Sq}(\mathrm{sg}(\mathsf{A}F, c)) \;\&\; \mathrm{sg}(\mathsf{A}F, c) \in \Delta_1 F$$

hence by Df 1.5 it only remains to prove, upon the premise $\mathrm{Prog}(F, C)$,

[4] $$\mathrm{Od}(c) \rightarrow (\mathsf{A}F)`c = F`\mathrm{sg}(\mathsf{A}F, c).$$

This goes by means of 1.8. Its application is enabled by the formula

$$\mathrm{Od}(c) \cdot\rightarrow\cdot (u)(u \in c' \rightarrow \mathrm{sg}(\mathsf{A}F, c')`u = (\mathsf{A}F)`u$$

which together with

$$(u)(\mathrm{Od}(u) \rightarrow \mathrm{Adp}(u, (\mathsf{A}F)`u, F)$$

gives

$$\mathrm{Od}(c) \cdot\rightarrow\cdot (u)(u \in c' \rightarrow \mathrm{Adp}(u, \mathrm{sg}(\mathsf{A}F, c')`u, F).$$

1.8 then gives

$$\mathrm{Od}(c) \rightarrow \mathrm{Adp}(\mathrm{sg}(\mathsf{A}F, c'), F),$$

hence

$$\begin{aligned} \mathrm{Od}(c) \quad \rightarrow \quad \mathrm{sg}(\mathsf{A}F, c')`c &= F`\mathrm{sg}(\mathrm{sg}(\mathsf{A}F, c'), c) \\ &= F`\mathrm{sg}(\mathsf{A}F, c) \qquad \text{by } 1.2, \end{aligned}$$

so that [4] results. So our proof of 1.6a is completed. Moreover it is arranged in such a way that the equivalence of 1.6a and 1.6b directly appears.

At the same time 1.5 together with 1.4a yield the formula

1.10 $$\mathrm{Prog}(F,C) \rightarrow \mathrm{Sq}(A) \;\&\; \Delta_1 A \equiv \{x \mid \mathrm{Od}(x)\} \;\&\; \mathrm{Adp}(A, F) \rightarrow A \equiv \mathsf{A}F$$

expressing that under the condition $\mathrm{Prog}(F, C)$ there is only one sequential class with the domain $\{x \mid \mathrm{Od}(x)\}$ which is adapted to F.

IV, § 2. The schema of transfinite recursion

The general recursion theorem is a principle means for proving in higher set theory the existence of functions with the domain $\{x \mid \mathrm{Od}(x)\}$. In particular in ordinal and cardinal arithmetic it yields the method of introducing *ordinal functions*, i.e. functions with ordinals as arguments and values, by means of transfinite recursion. This method here comes out to a transfinite iteration and we can formalize it by extending the iterator of III, § 4 to a *transfinite iterator*. We define this iterator by means of the adaptor:

Df 2.1 $\mathrm{I}(G, a) \equiv \mathsf{A}\{uv \mid \mathrm{Sq}(u) \mathbin{\&} \Delta_2 u^* \subseteq \Delta_1 G \mathbin{\&} (u=0 \rightarrow v=a) \mathbin{\&}$
$\mathbin{\&} (y)(\Delta_1 u = y' \rightarrow v = G`(u`y)) \mathbin{\&} (\mathrm{Lim}(\Delta_1 u) \rightarrow v = \sum_z(\Delta_1 u, u`z))\}$.

To this operator $\mathrm{I}(G, a)$ the general recursion theorem can directly be applied. Namely denoting the argument of A by $\mathfrak{F}(G, a)$, we have upon the premise

$$a \in C \mathbin{\&} \mathrm{Ft}(G) \mathbin{\&} \Delta_1 G \equiv C \mathbin{\&} \Delta_2 G \subseteq C \mathbin{\&} (z)(z \subseteq C \rightarrow \textstyle\sum z \in C)$$

the formula

$$\mathrm{Prog}(\mathfrak{F}(G, a), C)$$

and therefore by 1.6b

$$\mathrm{Sq}(\mathrm{I}(G, a)) \mathbin{\&} \Delta_1(\mathrm{I}(G, a)) \equiv \{x \mid \mathrm{Od}(x)\} \mathbin{\&} \Delta_2(\mathrm{I}(G, a)) \subseteq C \mathbin{\&}$$
$$\mathbin{\&} (x)(\mathrm{Od}(x) \rightarrow \mathrm{I}(G, a)`x = \mathfrak{F}(G, a)`\mathrm{sg}(\mathrm{I}(G, a), x)).$$

Thus we obtain

2.1 $a \in C \mathbin{\&} \mathrm{Ft}(G) \mathbin{\&} \Delta_1 G \equiv C \mathbin{\&} \Delta_2 G \subseteq C \mathbin{\&} (z)(z \subseteq C \rightarrow \sum z \in C) \mathbin{\&}$
$\mathbin{\&} H \equiv \mathrm{I}(G,a) \cdot\!\rightarrow\!\cdot \mathrm{Sq}(H) \mathbin{\&} \Delta_1 H \equiv \{x \mid \mathrm{Od}(x)\} \mathbin{\&} \Delta_2 H \subseteq C \mathbin{\&}$
$\mathbin{\&} H`0 = a \mathbin{\&} (x)(\mathrm{Od}(x) \rightarrow H`x' = G`(H`x)) \mathbin{\&} (x)(\mathrm{Lim}(x) \rightarrow$
$\rightarrow H`x = \sum_z(x, H`z))$.

The comparison of this formula with III, 4.1 shows that the operator I is indeed an extension of the operator J.

For the ordinal theory we have to apply the iterator $\mathrm{I}(G,a)$ to functions G with the domain $\{x \mid \mathrm{Od}(x)\}$. In these applications we need not verify in 2.1 the premise on C

$$(z)(z \subseteq C \rightarrow \textstyle\sum z \in C)$$

since it is clearly satisfied for C being $\{x \mid \mathrm{Od}(x)\}$.

We now can define the *ordinal arithmetic functions* sum, product and exponentiation by means of I as we did it for the corresponding numeral functions by means of the operator J in III, Df 4.4–Df 4.6.

In order not to deviate from the familiar notations we employ here the same function signs for the ordinal functions as we used for the corresponding numeral functions. This is here to be understood in the way that in the frame of full ordinal theory (in distinction from number theory) the three function signs are to be used according to the now following definitions:

Df 2.2 $$a+b=\mathrm{I}(\{xy \mid y=x'\}, a)`b$$

Df 2.3 $$a \cdot b=\mathrm{I}(\{xy \mid y=x+a\}, 0)`b$$

Df 2.4 $$a^b=\mathrm{I}(\{xy \mid y=x \cdot a\}, 0')`b.$$

From these definitions by 2.1 we get

2.2 $$\mathrm{Od}(a) \,\&\, \mathrm{Od}(b) \rightarrow \mathrm{Od}(a+b) \,\&\, \mathrm{Od}(a \cdot b) \,\&\, \mathrm{Od}(a^b),$$

as also the recursive laws [1])

2.3 $$\begin{cases} \mathrm{Od}(a) \rightarrow a+0=a \\ \mathrm{Od}(a) \,\&\, \mathrm{Od}(b) \rightarrow a+b'=(a+b)' \\ \mathrm{Od}(a) \,\&\, \mathrm{Lim}(b) \rightarrow a+b=\sum_x(b, a+x) \end{cases}$$

2.4 $$\begin{cases} \mathrm{Od}(a) \rightarrow a \cdot 0=0 \\ \mathrm{Od}(a) \,\&\, \mathrm{Od}(b) \rightarrow a \cdot b'=a \cdot b+a \\ \mathrm{Od}(a) \,\&\, \mathrm{Lim}(b) \rightarrow a \cdot b=\sum_x(b, a \cdot x) \end{cases}$$

2.5 $$\begin{cases} \mathrm{Od}(a) \rightarrow a^0=0' \\ \mathrm{Od}(a) \,\&\, \mathrm{Od}(b) \rightarrow a^{b'}=a^b \cdot a \\ \mathrm{Od}(a) \,\&\, \mathrm{Lim}(b) \rightarrow a^b=\sum_x(b, a^x). \end{cases}$$

From these formulas evidently follows that the ordinal functions $a+b$, $a \cdot b$, a^b, have for numeral arguments the same values as the corresponding numeral functions, so that the ordinal functions can figure in the domain of natural numbers instead of the numeral functions.

---

[1]) For the ordinal arithmetic functions we use the familar conventions for sparing parentheses.

By means of 2.3–2.5 we can derive by transfinite induction the formulas expressing the computation laws of the ordinal arithmetic functions:

$$2.6\quad \left\{\begin{array}{ll} \mathrm{Od}(a)\ \&\ \mathrm{Od}(b)\ \&\ \mathrm{Od}(c) & \rightarrow a+(b+c)=(a+b)+c \\ & \rightarrow a\cdot(b\cdot c)=(a\cdot b)\cdot c \\ & \rightarrow a\cdot(b+c)=a\cdot b+a\cdot c \\ & \rightarrow a^{b+c}=a^b\cdot a^c \\ & \rightarrow a^{b\cdot c}=(a^b)^c. \end{array}\right.$$

The other computation laws of ordinary arithmetic are not generally valable for ordinals.

From 2.1 we also can pass to the schema of ordinal recursion, which applies in the following form:

2.7 Let $\mathfrak{t}(c)$ and $\mathfrak{p}(a,b)$ be terms (where the indicated arguments need not all occur) and $\mathfrak{A}(c)$ some formula, and the following conditions be satisfied

$$(C)\quad \begin{array}{l} \mathfrak{A}(a) \rightarrow \mathrm{Od}(\mathfrak{t}(a)) \\ \mathrm{Od}(n)\ \&\ \mathfrak{A}(a) \rightarrow \mathrm{Od}(\mathfrak{p}(n,a)), \end{array}$$

then a function symbol $\mathfrak{f}(a, n)$, can be introduced with the formulas

[1] $\mathfrak{A}(a) \rightarrow \mathfrak{f}(a, 0)=\mathfrak{t}(a)$

[2] $\mathfrak{A}(a)\ \&\ \mathrm{Od}(n) \rightarrow \mathrm{Od}(\mathfrak{f}(a, n))\ \&\ \mathfrak{f}(a, n')=\mathfrak{p}(\mathfrak{f}(a, n), a)$

[3] $\mathfrak{A}(a)\ \&\ \mathrm{Lim}(b) \rightarrow \mathfrak{f}(a, b)=\sum_{\mathfrak{x}}(b, \mathfrak{f}(a, \mathfrak{x}))$

constituting its recursive definition.

Our schema in particular includes the case where $\mathfrak{A}(c)$ is $\mathrm{Od}(c)$, which gives the recursive definition of ordinal functions with two arguments, one of which has the rôle of a parameter. But also the case of more parameters is included by taking for $\mathfrak{A}(a)$ the predicate of a being a pair, or more generally a $\mathfrak{k}$-tuplet of ordinals.

The derivability of the schema 2.7 is with the help of the transfinite iterator. Namely defining

[4] $\mathfrak{f}(a, c)=\mathrm{I}(\{\mathfrak{x}\mathfrak{y} \mid \mathrm{Od}(\mathfrak{x})\ \&\ \mathfrak{y}=\mathfrak{p}(\mathfrak{x}, a)\}), \mathfrak{t}(a))\text{`}c,$

we get upon the condition (C) by 2.1 the formulas [1], [2], and [3].

A case often present in the applications of the transfinite iterator or of the schema 2.7 is that of introduction of *Normalfunktionen*, i.e. of ordinal functions which are *strictly monotonic* and *continuous*.

An ordinal function F is called strictly monotonic if

$$\mathrm{Od}(b)\ \&\ a \in b \rightarrow F`a \in F`b,$$

and it is called continuous, if

$$\mathrm{Lim}(b) \rightarrow F`b = \sum_x (b, F`x).$$

Thus the formal definition of a Normalfunktion is

Df 2.5 $\mathrm{Nft}(F) \leftrightarrow \mathrm{Sq}(F)\ \&\ \Delta_1 F \equiv \{x \mid \mathrm{Od}(x)\}\ \&\ \Delta_2 F \subseteq \{x \mid \mathrm{Od}(x)\}\ \&$
$\&\ (x)(y)(\mathrm{Od}(y)\ \&\ x \in y \rightarrow F`x \in F`y)\ \&\ (z)(\mathrm{Lim}(z) \rightarrow F`z = \sum_x (z, F`x)).$

In a wider sense the terminus "Normalfunktion" is applied in set theory also for segments of Normalfunktionen in the defined sense.

Normalfunktionen are obtained from 2.7 in the case that besides the conditions (C) also the condition

(C_1) $$\mathrm{Od}(n)\ \&\ \mathfrak{A}(a) \rightarrow n \in \mathfrak{p}(n, a)$$

is fulfilled. Indeed then the function $\mathfrak{f}(a, c)$ defined by [4] is with respect to c for any a satisfying $\mathfrak{A}(a)$ a Normalfunktion. Namely the property of continuity holds already by [3] and that of strict monotonity can be proved to hold from (C_1) using transfinite induction.

By this way we immediately recognize that

2.8 $$\begin{cases} \mathrm{Od}(a) \rightarrow \mathrm{Nft}(\{xy \mid \mathrm{Od}(x)\ \&\ a + x = y\}), \\ \mathrm{Od}(a)\ \&\ 0 \in a \rightarrow \mathrm{Nft}(\{xy \mid \mathrm{Od}(x)\ \&\ a \cdot x = y\}), \\ \mathrm{Od}(a)\ \&\ 0' \in a \rightarrow \mathrm{Nft}(\{xy \mid \mathrm{Od}(x)\ \&\ a^x = y\}). \end{cases}$$

By specializing the application of 2.7 to the case where $\mathfrak{t}(a)$ is a, $\mathfrak{A}$ is Od and $\mathfrak{p}$ has only one argument, so that instead of (C), (C_1) we simply have $\mathrm{Od}(n) \rightarrow \mathrm{Od}(\mathfrak{p}(n))\ \&\ n \in \mathfrak{p}(n)$, we obtain

2.9 $$\mathrm{Od}(a)\ \&\ (\mathfrak{x})\, \mathrm{Od}(\mathfrak{x}) \rightarrow (\mathrm{Od}(\mathfrak{p}(\mathfrak{x}))\ \&\ \mathfrak{x} \in \mathfrak{p}(\mathfrak{x})) \rightarrow$$
$$\rightarrow \mathrm{Nft}(\mathrm{I}(\{\mathfrak{x}\mathfrak{y} \mid \mathrm{Od}(\mathfrak{x})\ \&\ \mathfrak{y} = \mathfrak{p}(\mathfrak{x})\}, a)).$$

Of the general properties of Normalfunktionen there are in particular those following from strict monotonity. Indeed every strict monotonic ordinal function is a one-to-one correspondence and its converse function is again monotonic; further every value of such a function is at least as high as the argument. The first statement results from III, 1.11, the second by using the principle of least ordinal number in connection with III, 2.12, 2.8 and 2.14. Applying these statements in particular to Normalfunktionen, we have

2.10 $\mathrm{Nft}(F) \rightarrow \mathrm{Crs}(F) \,\&\, (x)(y)(\mathrm{Od}(x) \,\&\, \mathrm{Od}(y) \cdot\!\!\rightarrow\!\!\cdot\, F`x \in F`y \rightarrow x \in y)$

2.11 $\mathrm{Nft}(F) \cdot\!\!\rightarrow\!\!\cdot (x)(\mathrm{Od}(x) \rightarrow x \in F`x \vee x = F`x).$

The ordinals n such that $n = F`n$ are called *critical points* of F. The existence of critical points for any Normalfunktion will be provable only when we shall have the axiom of infinity available.

IV, § 3. Generated numeration

A particular noticeable application of the general recursion theorem is to numerations of sets. By a *numeration* of a set c we understand a one-to-one correspondence between an ordinal and c. In a numeration of c the elements of c follow one another like the elements of an ordinal in their order given by the $\in$-relation [1]).

The formal definition is:

Df 3.1 $\mathrm{Num}(s, c) \leftrightarrow \mathrm{Sq}(s) \,\&\, \mathrm{Crs}(s) \,\&\, \Delta_2 s = c.$

Concerning this concept we draw from the general recursion theorem a theorem of *generated numeration*. We generally call a sequence s generated by a function G if $s`x$ is the value assigned by G to the converse domain of the x-segment of s; formally

Df 3.2 $\mathrm{Gen}(s, G) \leftrightarrow \mathrm{Sq}(s) \,\&\, (x)(x \in \Delta_1 s \rightarrow s`x = G`\Delta_2 \mathrm{sg}(s, x)).$

Now the assertion of the said theorem is that for any function

---

[1]) The order of the elements of a set which comes about by a numeration will be explicitly discussed in the next chapter, § 2.

G which assigns to every proper subset p of c an element of $c^{-}p$ there exists a numeration of c generated by G:

3.1 $$\mathrm{Ft}(G)\ \&\ (z)(z \subset c \to z \in \Delta_1 G\ \&\ G`z \in c^{-}z) \to \\ \to (Ex)(\mathrm{Num}(x, c)\ \&\ \mathrm{Gen}(x, G)).$$

For the proof [1]) we first state that for any function G satisfying the premise of 3.1 the function defined by the expression

[1] $$\{xy \mid \mathrm{Sq}(x)\ \&\ (\Delta_2 x \subset c\ \&\ y = G`\Delta_2 x \cdot \mathbf{v} \cdot \Delta_2 x = c\ \&\ y = G`0)\}$$

– let us denote it by $\mathfrak{F}(G, c)$ – has the property $\mathrm{Prog}(\mathfrak{F}(G, c), c^*)$. Thus we can apply to it 1.6; substituting c^* for C and $\mathfrak{F}(G, c)$ for F. So we obtain – still indicating briefly the class expression $\mathsf{A}\mathfrak{F}(G, c)$ by $\mathfrak{H}$

[2] $$\mathrm{Adp}(\mathfrak{H}, \mathfrak{F}(G, c))\ \&\ \Delta_1 \mathfrak{H} \equiv \{x \mid \mathrm{Od}(x)\}\ \&\ \Delta_2 \mathfrak{H} \subseteq c^*$$

and by the definitions Df 1.5, Df 1.6, Df 1.8 and by [1]

[3] $$\mathrm{Od}(n) \to \Delta_2 \mathrm{sg}(\mathfrak{H}, n) \subset c\ \&\ \mathfrak{H}`n = G`\Delta_2 \mathrm{sg}(\mathfrak{H}, n) \cdot \mathbf{v} \cdot \Delta_2 \mathrm{sg}(\mathfrak{H}, n) = c\ \& \\ \&\ \mathfrak{H}`n = G`0).$$

Now this function $\mathfrak{H}$ cannot be a one-to-one correspondence; for its converse domain is a subclass of c and thus represented by a set, and so by application of II, 3.8 to $\breve{\mathfrak{H}}$ it would follow that the class $\{x \mid \mathrm{Od}(x)\}$ is represented by a set what, as we know, does not hold.

Therefore by the principle of the least ordinal number there must exist a least ordinal l such that for some ordinal $m \in l$ we have $\mathfrak{H}`l = \mathfrak{H}`m$. Hence denoting by $\mathfrak{l}$ the term

[4] $$\mu_x(\mathrm{Od}(x)\ \&\ (Ez)(z \in x\ \&\ \mathfrak{H}`z = \mathfrak{H}`x))$$

we have

[5] $$\mathrm{Crs}(\mathrm{sg}(\mathfrak{H}, \mathfrak{l}))\ \&\ \overline{\mathrm{Crs}}(\mathrm{sg}(\mathfrak{H}, \mathfrak{l}'))).$$

Now at all events, by [2], $\Delta_2 \mathrm{sg}(\mathfrak{H}, \mathfrak{l}) \subseteq c$. But we cannot have $\Delta_2 \mathrm{sg}(\mathfrak{H}, \mathfrak{l}) \subset c$; for then it would follow by [3]

$$\mathfrak{H}`\mathfrak{l} = G`\Delta_2 \mathrm{sg}(\mathfrak{H}, \mathfrak{l})$$

and by the premise of 3.1

$$G`\Delta_2 \mathrm{sg}(\mathfrak{H}, \mathfrak{l}) \in c^{-}\Delta_2 \mathrm{sg}(\mathfrak{H}, \mathfrak{l}),$$

---

[1]) In the following it is assumed $c \neq 0$; in the other case 3.1 is obvious.

which entails

$$\mathrm{Crs}(\mathrm{sg}(\mathfrak{H}, \mathfrak{l}); \langle \mathfrak{l}', \mathfrak{H}\text{‘}\mathfrak{l}\rangle),$$

i.e. $\mathrm{Crs}(\mathrm{sg}(\mathfrak{H}, \mathfrak{l}'))$, whereas by [5] $\overline{\mathrm{Crs}}(\mathrm{sg}(\mathfrak{H}, \mathfrak{l}'))$. Thus $\Delta_2\mathrm{sg}(\mathfrak{H}, \mathfrak{l}) = c$, and denoting $\mathrm{sg}(\mathfrak{H}, \mathfrak{l})$ by $\mathfrak{s}$ we have according to Df 3.1 and Df 3.2, using also [3] and 1.2,

$$\mathrm{Num}(\mathfrak{s}, c) \;\&\; \mathrm{Gen}(\mathfrak{s}, G).$$

Hence it results that upon the premise of the theorem of generated numeration 3.1 its existential assertion is fulfilled by the term $\mathrm{sg}(\mathsf{A}\mathfrak{F}(G, c), \mathfrak{l})$, where $\mathfrak{F}(G, c)$ is the class term [1] and $\mathfrak{l}$ the set term [4].

There is no difficulty to prove also by means of the principle of the least ordinal number the formula

[6] $$\mathrm{Num}(s, c) \;\&\; \mathrm{Gen}(s, G) \rightarrow s = \mathrm{sg}(\mathsf{A}\mathfrak{F}(G, c), \mathfrak{l}),$$

which expresses that for a given c and G the numeration of c generated by G is uniquely determined.

In order to infer from the theorem 3.1 the existence of a numeration of a set the only thing required is that we can prove for any set c the existence of a function G with the property

[7] $$(z)(z \subset c \rightarrow z \in \Delta_1 G \;\&\; G\text{‘}z \in c^- z).$$

This will become possible later on, when we shall have available the axiom of choice.

The proof of 3.1 can also be given by the method of Zermelo's first demonstration of the wellorder theorem [1904], without applying the general recursion theorem. This goes as follows:

We consider the class N of numerations of subsets of c generated by G. Of course there exist such numerations. Let s and t be elements of N; then for every ordinal k which is in the domain of both we have $s\text{‘}k = t\text{‘}k$. For, otherwise there would be among the ordinals k such that $s\text{‘}k \neq t\text{‘}k$ a lowest one l. But then we should have $\mathrm{sg}(s, l) = \mathrm{sg}(t, l)$ and this would yield a contradiction between the defining property Df 3.2 of a numeration generated by G and our assumption about l.

Therefore the class $\bigcup N$ is a function; and further, if $\langle m, b\rangle$ is an element of this function, then for every element r of N such that $m \in \Delta_1 r$ we have $r`m = b$. From this in particular follows that $\bigcup N$ is a one-to-one correspondence, since every element of N is a one-to-one correspondence. Moreover $\Delta_2 \bigcup N \subseteq c^*$, hence $\mathrm{Rp}(\Delta_2 \bigcup N)$ and therefore $\mathrm{Rp}(\Delta_1 \bigcup N)$ and $\mathrm{Rp}(\bigcup N)$. Thus we have proved

3.2 $\quad \mathrm{Rp}(\bigcup\{x \mid (Ey)(y \subseteq c \;\&\; \mathrm{Num}(x, y) \;\&\; \mathrm{Gen}(x, G))\})$.

Let now d be the set representing $\bigcup N$, then $\Delta_1 d$ is a transitive set of ordinals and thus by III, 1.14 itself an ordinal. So d is a numeration of a subset of c. Besides it is generated by G. For, if $m \in \Delta_1 d$ then m is in the domain of some element r of N and for every ordinal k not higher than m we have $d`k = r`k$.

The only thing still to be proved is that $\Delta_2 d = c$. But in the other case the set $d; \langle \Delta_1 d, G`\Delta_2 d\rangle$ would be an element of $\bigcup N$, hence $\langle \Delta_1 d, G`\Delta_2 d\rangle \in d$, and we should have $\Delta_1 d \in \Delta_1 d$. –

Still by means of the adaptor we can set up, for any class of ordinals C such that there is for every ordinal a higher one in C, a monotonic one-to-one mapping of $\{x \mid \mathrm{Od}(x)\}$ onto C.

For this we take the function $\mathfrak{F}$ given by the expression

[1] $\quad \{xy \mid \mathrm{Sq}(x) \;\&\; \Delta_2 x^* \subseteq C \;\&\; y = \mu_z(z \in C \;\&\; z \notin \Delta_2 x)\}$.

$\mathfrak{F}$ fulfils the condition $\mathrm{Prog}(\mathfrak{F}, C)$. Namely for every C-sequence s there is by our assumption on C an element in C, higher than $\sum \Delta_2 s$ and hence not in $\Delta_2 s$. By this also follows that for every C-sequence s we have

[2] $$\mathfrak{F}`s \notin \Delta_2 s$$

[3] $$k \in C \;\&\; k \notin \Delta_2 s \rightarrow \mathfrak{F}`s \subseteq k.$$

Taking now the adaptor $\mathsf{A}\mathfrak{F}$ we have by 1.6b

[4] $$\mathrm{Ft}(\mathsf{A}\mathfrak{F})$$

[5] $$\Delta_1 \mathsf{A}\mathfrak{F} \equiv \{x \mid \mathrm{Od}(x)\}$$

[6] $$\Delta_2 \mathsf{A}\mathfrak{F} \subseteq C$$

[7] $$\mathrm{Od}(a) \rightarrow (\mathsf{A}\mathfrak{F})`a = \mathfrak{F}`(\mathrm{sg}(\mathsf{A}\mathfrak{F}, a)),$$

and besides by [2], [3] and [7]

[8] $\quad \mathrm{Od}(a) \rightarrow (\mathsf{A}\mathfrak{F})`a \notin \Delta_2 \mathrm{sg}(\mathsf{A}\mathfrak{F}, a)$

[9] $\quad \mathrm{Od}(a) \cdot\!\!\rightarrow\!\cdot\, k \in C \;\&\; k \notin \Delta_2 \mathrm{sg}(\mathsf{A}\mathfrak{F}, a) \rightarrow (\mathsf{A}\mathfrak{F})`a \subseteq k.$

Our statement will be proved if we show first

$$\mathrm{Od}(b) \;\&\; a \in b \rightarrow (\mathsf{A}\mathfrak{F})\text{'}a \in (\mathsf{A}\mathfrak{F})\text{'}b,$$

and secondly

$$C \subseteq \Delta_2 \mathsf{A}\mathfrak{F}.$$

The first formula results from

$$\begin{aligned} \mathrm{Od}(b) \;\&\; a \in b &\rightarrow \Delta_2\mathrm{sg}(\mathsf{A}\mathfrak{F}, a) \subseteq \Delta_2\mathrm{sg}(\mathsf{A}\mathfrak{F}, b) \\ &\rightarrow (\mathsf{A}\mathfrak{F})\text{'}b \notin \Delta_2\mathrm{sg}(\mathsf{A}\mathfrak{F}, a) && \text{by [8]} \\ &\rightarrow (\mathsf{A}\mathfrak{F})\text{'}a \subseteq (\mathsf{A}\mathfrak{F})\text{'}b, && \text{by [6] and [9],} \end{aligned}$$

and

$$\begin{aligned} \mathrm{Od}(b) \;\&\; a \in b &\rightarrow (\mathsf{A}\mathfrak{F})\text{'}a \in \Delta_2\mathrm{sg}(\mathsf{A}\mathfrak{F}, b) \\ &\rightarrow (\mathsf{A}\mathfrak{F})\text{'}b \notin \Delta_2\mathrm{sg}(\mathsf{A}\mathfrak{F}, b) && \text{by [8]} \\ &\rightarrow (\mathsf{A}\mathfrak{F})\text{'}a \neq (\mathsf{A}\mathfrak{F})\text{'}b. \end{aligned}$$

The second formula can be strengthened to

[10] $$b \in C \rightarrow (Ex)(\mathrm{Od}(x) \;\&\; x \subseteq b \;\&\; (\mathsf{A}\mathfrak{F})\text{'}x = b).$$

This we prove by transfinite induction with respect to b after adding the redundant premise Od(b). By the induction premise we have

$$\mathrm{Od}(b) \;\&\; a \in b \;\&\; a \in C \rightarrow a \in \Delta_2\mathrm{sg}(\mathsf{A}\mathfrak{F}, b)$$

and thus by contraposition

[11] $$\mathrm{Od}(b) \;\&\; a \in C \;\&\; a \notin \Delta_2\mathrm{sg}(\mathsf{A}\mathfrak{F}, b) \rightarrow b \subseteq a.$$

Now

$$b \in \Delta_2\mathrm{sg}(\mathsf{A}\mathfrak{F}, b) \rightarrow (Ex)(x \in b \;\&\; (\mathsf{A}\mathfrak{F})\text{'}x = b)$$

$$\begin{aligned} b \in C \;\&\; b \notin \Delta_2\mathrm{sg}(\mathsf{A}\mathfrak{F}, b) &\rightarrow b = \mu_z(z \in C \;\&\; z \notin \Delta_2\mathrm{sg}(\mathsf{A}\mathfrak{F}, b)) \\ &\rightarrow b = (\mathsf{A}\mathfrak{F})\text{'}b, && \text{by [7] and [1]} \end{aligned}$$

so that [10] results.

On the whole we have obtained the theorem

3.3 $$\begin{aligned} C \subseteq \{x \mid \mathrm{Od}(x)\} \;\&\; (z)(\mathrm{Od}(z) \rightarrow (Ey)(y \in C \;\&\; z \in y)\cdot\!\rightarrow\!\cdot \\ \cdot\!\rightarrow\!\cdot \{x \mid \mathrm{Od}(x)\} \ulcorner_{\mathsf{A}\mathfrak{F}}\urcorner\, C \;\; \& \\ \& \;\; (\mathrm{Od}(b) \;\&\; a \in b \rightarrow (\mathsf{A}\mathfrak{F})\text{'}a \in (\mathsf{A}\mathfrak{F})\text{'}b), \end{aligned}$$

with $\mathfrak{F}$ being the class term [1].

CHAPTER V

POWER; ORDER; WELLORDER

V, § 1. Comparison of powers

The statements about the equivalence of sets constitute the starting point for the comparison of powers to which we are lead by the following definitions. A set a is called of *equal power* as b if $a \sim b$, of *at most equal power* as b if a is equivalent to a subset of b, of *lower power* than b (or b of *higher power* than a) if a is of at most equal power as b but not of equal power as b.

Formally we express the last two definitions by

Df 1.1 $$a \preceq b \leftrightarrow (Ex)(x \subseteq b \;\&\; a \sim x)$$

Df 1.2 $$a \prec b \leftrightarrow a \preceq b \;\&\; \overline{a \sim b}.$$

Of course we have

1.1 $$a \preceq a, \quad \overline{a \prec a},$$

and

1.2 $$\begin{array}{ll} a \preceq c \;\&\; a \sim b \rightarrow b \preceq c, & a \preceq c \;\&\; c \sim b \rightarrow a \preceq b, \\ a \prec c \;\&\; a \sim b \rightarrow b \prec c, & a \prec c \;\&\; c \sim b \rightarrow a \prec b. \end{array}$$

The three formulas 1.2 are inferred from the composibility of one-to-one correspondences. Likewise by using this composibility the proof of the *equivalence theorem*

1.3 $$a \preceq b \;\&\; b \preceq a \rightarrow a \sim b$$

is reduced to that of the *Bernstein-Schröder theorem*

1.4 $$a \subseteq b \;\&\; b \subseteq c \;\&\; a \sim c \rightarrow b \sim c.$$

This theorem follows from a corresponding statement on classes. Namely we can set up a class term $\mathfrak{K}$ (with the three variables B, C, P) such that the formula

1.5 $$A \subseteq B \;\&\; B \subseteq C \;\&\; C \underset{P}{\sim} A \rightarrow C \underset{\mathfrak{K}}{\sim} B$$

is provable. Indeed if $\mathfrak{A}(c)$ is the predicate expression

$$(Ez)(z \in C \cap \bar{B} \ \& \ c \in \Delta_2 \mathrm{J}(P, z))$$

and $\mathfrak{R}$ the class term

$$\{xy \mid \mathfrak{A}(x) \ \& \ y = P^{\iota}x \cdot \mathbf{v} \cdot \overline{\mathfrak{A}}(x) \ \& \ y = x \ \& \ x \in C\}$$

then first from the expression of $\mathfrak{R}$ we get

[1] $$\mathrm{Ft}(\mathfrak{R}) \ \& \ \Delta_1 \mathfrak{R} \equiv C.$$

Further we have

$$a \in C \rightarrow P^{\iota}a \in A, \quad a \in C \cap \bar{B} \rightarrow \mathfrak{A}(a)$$

and thus

$$a \in C \rightarrow P^{\iota}a \in B, \quad a \in C \ \& \ \overline{\mathfrak{A}}(a) \rightarrow a \in B,$$

which together give

[2] $$\Delta_2 \mathfrak{R} \subseteq B.$$

Moreover we have

$$\mathfrak{A}(a) \ \& \ a \in B \rightarrow (Ex)(\mathfrak{A}(x) \ \& \ a = P^{\iota}x),$$

which yields, together with the expression of $\mathfrak{R}$,

[3] $$B \subseteq \Delta_2 \mathfrak{R}.$$

Finally, by means of

$$\mathfrak{A}(a) \ \& \ b = P^{\iota}a \rightarrow \mathfrak{A}(b)$$

and

$$\mathrm{Crs}(P)$$

we get

[4] $$\mathrm{Crs}(\mathfrak{R}).$$

Thus we obtain, combining [1]–[4], the formula 1.5.

Moreover we get from 1.1, 1.2, 1.3

1.6 $$a \prec b \ \& \ b \prec c \rightarrow a \prec c, \ a \prec b \rightarrow \overline{b \prec a}.$$

As a simple consequence of 1.3 we still note

1.7 $$a \prec b \rightarrow \overline{b \subseteq a}.$$

For a satisfactory theory of power still two things are lacking.

First we have not the alternative

(A) $$a \preceq b \vee b \preceq a$$

which would express the comparability of sets with respect to power.

Further it is to be observed that besides the given usual definition of "at most equal power" there would be an other one likewise natural by saying that a non empty set a is of at most equal power as b if there exists a functional set mapping b onto a. In order that this definition do not conflict with our given one (Df 1.1), we ought to be able to derive the equivalence

(B) $$a \preceq b \leftrightarrow a=0 \vee (Ex)(\mathrm{Ft}(x) \,\&\, \Delta_1 x=b \,\&\, \Delta_2 x=a).$$

Here the implication from left to right is easily seen to be derivable. However for the inverse implication no proof is available.

This inverse implication, which it is obviously sufficient to prove without the member $a=0$, so that it becomes

(C) $$(Ex)(\mathrm{Ft}(x) \,\&\, \Delta_1 x=b \,\&\, \Delta_2 x=a) \rightarrow a \preceq b,$$

would almost directly result if we had at our disposal the formula

(D) $$\mathrm{Ft}(c) \,\&\, \Delta_2 c=a \rightarrow (Ex)(\mathrm{Ft}(x) \,\&\, x \subseteq \breve{c} \,\&\, \Delta_1 x=a)$$

which expresses the assertion that for any functional set there exists an inverse functional set. The passage from (D) to (C) goes by the aid of the formulas

$$\Delta_1 c=b \,\&\, h \subseteq \breve{c} \rightarrow \Delta_2 h \subseteq b$$

$$\mathrm{Ft}(c) \,\&\, h \subseteq \breve{c} \,\&\, \mathrm{Ft}(h) \rightarrow \mathrm{Crs}(h).$$

The formula (D) however can be seen to be equivalent with the following one, appearingly somewhat more general

(E) $$\mathrm{Ps}(c) \rightarrow (Ex)(x \subseteq c \,\&\, \Delta_1 s=\Delta_1 c \,\&\, \mathrm{Ft}(x))$$

which is one of the forms of stating the axiom of choice. We leave the proof of this equivalence (in the sense of equal deducibility) for the next chapter, where we shall discuss the axiom of choice. There we shall also see that with the aid of the axiom of choice the formula (A) becomes derivable.

Another feature present in Cantor's theory of power which does not yet result in our frame is that to every power exists a higher one. This in Cantor's theory follows from the theorem that the set of the subsets of a set a is of higher power than a. For this theorem – already in its statement – the assumption is essential that there exists for every set a the *set* of its subsets. This existential assertion cannot be derived from our present frame – as was shown in [Bernays 1948]. In other words we here only know that there exists for a set a the *class* of its subsets $\{x \mid x \subseteq a\}$, but not yet that this class is represented by a set. The assertion of this representation is the contents of the potency axiom which we shall introduce in the next chapter.

At all events in the present frame Cantor's method of proof of his just mentioned theorem can be used to prove that there exists for a set a no function mapping a^* onto $\{x \mid x \subseteq a\}$, so that a fortiori there exists no one-to-one correspondence between $\{x \mid x \subseteq a\}$ and a^* or a subset of a^*. The proof will be given in a positive form by deriving for any function F mapping a^* onto a class C of subsets of a the existence of a subset of a not belonging to C. For the proof we do not even use that $C \subseteq \{x \mid x \subseteq a\}$; thus, what we are to prove is

1.8 $\quad \mathrm{Ft}(F) \,\&\, \Delta_1 F \equiv a^* \,\&\, \Delta_2 F \equiv C \rightarrow (Ez)(z \subseteq a \,\&\, z \notin C).$

Let us for this denote the set term $a \cap \{x \mid x \notin F\text{'}x\}$ by $\mathfrak{t}$. Then we have

[1] $$\mathfrak{t} \subseteq a$$

$$c \in \mathfrak{t} \leftrightarrow c \in a \,\&\, c \notin F\text{'}c$$

and thus also

$$c \in a \,\&\, \mathfrak{t} = F\text{'}c \rightarrow (c \in \mathfrak{t} \leftrightarrow c \notin \mathfrak{t})$$

hence

$$c \in a \rightarrow \mathfrak{t} \neq F\text{'}c.$$

Therefore we have

[2] $$\mathrm{Ft}(F) \,\&\, \Delta_1 F \equiv a^* \,\&\, \Delta_2 F \equiv C \rightarrow \mathfrak{t} \notin C.$$

From [1] and [2] results 1.8.

On the other hand there exists a trivial mapping of a^* onto a subclass of $\{x \mid x \subseteq a\}$, namely

1.9 $$a^* \ulcorner_{\{xy \mid y=[x]\}}\urcorner \{x \mid (Ez)(z \in a \;\&\; x=[z]\}.$$

So we get here the relations stated by Cantor's mentioned theorem roughly expressed by saying that a set has more subsets than it has elements, – only that the comparison by our derivation is not between a and a set of its subsets but rather between a and the class of its subsets.

Still it is to be observed that a corresponding statement about a class A instead of a set a, namely that there cannot be a one-to-one correspondence between A and the class of its subsets, does not generally hold. Indeed for the class V we have

1.10 $$V \ulcorner_{\{xy \mid x=y\}}\urcorner \{x \mid x^* \subseteq V\}$$

since every set is an element and a subset of V. Here it appears that Cantor's paradox connected with the set of all sets is removed in our system by the distinction of sets and classes. Of course it results that the class V is not represented by a set.

V, § 2. Order and partial order

An order of a set or a class comes about by an ordering relation with certain formal properties. Extensionally an order is characterized by a class C whose elements are the pairs $\langle a, b\rangle$ of elements of the set or class in question such that either a equal b or a precedes b in the order. Of course the class C is not an arbitrary class of pairs but must have the formal properties corresponding to those of an ordering relation. Let us state these for the case of a class A to be ordered by defining the predicator $\mathrm{Or}(C, A)$ («C is an order of the class A»):

Df 2.1 $$\mathrm{Or}(C, A) \leftrightarrow \mathrm{Ps}(C) \;\&\; (x)(y)(\langle x, y\rangle \in C \leftrightarrow x \in A \;\& \;\& y \in A \;\&\; (x=y \vee \langle y, x\rangle \notin C)) \;\&\; (x)(y)(z)(\langle x, y\rangle \in C \;\& \;\& \langle y, z\rangle \in C \rightarrow \langle x, z\rangle \in C).$$

The definition of $\mathrm{Or}(C, a)$ is quite corresponding.

We immediately have

2.1 $$\mathrm{Or}(C, A) \rightarrow C \subseteq A \times A \ \& \ \Delta_1 C \equiv A \ \& \ \Delta_2 C \equiv A$$

and

$$\mathrm{Or}(C, a) \leftrightarrow \mathrm{Or}(C, a^*),$$

therefore also, by II 3.16

2.2 $$\mathrm{Or}(C, a) \rightarrow C \subseteq (a \times a)^*.$$

By the last formula it follows that an ordering class of a set a is always represented by a set, and so, defining

Df 2.2 $$\mathrm{Or}(c, a) \leftrightarrow \mathrm{Or}(c^*, a),$$

we have

2.3 $$\mathrm{Or}(C, a) \rightarrow (Ex)(x^* \equiv C \ \& \ \mathrm{Or}(x, a))$$ [1].

We also state

2.4 $$\mathrm{Or}(C, A) \ \& \ B \subseteq A \rightarrow \mathrm{Or}(C \cap (B \times B), B),$$

thus every order C of a class A determines an order of any subclass B of A; we call it the order C restricted to B.

Let us further define

Df 2.3 $$\mathrm{Or}(C) \leftrightarrow \mathrm{Or}(C, \Delta_1 C)$$

Df 2.4 $$\mathrm{Or}(c) \leftrightarrow \mathrm{Or}(c^*)$$

«C (or c) is an ordering class (or set)» or briefly «C (or c) is an order».

Then we have by Df 2.1 and Df 2.2

2.5 $$\mathrm{Or}(C, A) \rightarrow \mathrm{Or}(C), \quad \mathrm{Or}(c) \rightarrow (Ex)\mathrm{Or}(c, x).$$

By means of an ordering class the predicate "r before s" relative to elements of the class or set ordered by C is expressed by $\langle r, s \rangle \in C \ \& \ r \neq s$. We also speak of "the order of A (or of a) by the class C" when C is an ordering class of A (or of a).

It might be asked if it would not be simpler and more natural to define $\mathrm{Or}(C, a)$ in such a way that the elements of an order C

[1]) At speaking in the following of orders of a set a we always mean ordering sets of a. This gives in virtue of 2.2 no restriction on the order. In the symbolic formulations no ambiguity can arise.

of a are the pairs $\langle r, s \rangle$ of elements of a with r preceding s, thus excluding the pairs $\langle r, r \rangle$. But then we would have the disadvantage that an ordering class of an unit set would have to be empty and thus different unit sets, regarded as ordered sets, could not be distinguished.

There is still an other way of characterizing an order of a set a by a class [1]. Namely by an order of a those subsets of a are distinguished which constitute an *initial section* (Anfangsstück), i.e. which satisfy the condition that if c is an element of it and b an element of a which precedes c by the order, then b is an element of it. In a corresponding way the concept of a *terminal section* (Endstück) can be defined. Now it is possible to characterize an order of a set by the class of the initial sections (relative to the order), or that of the terminal sections, or else by a suitable subclass of one of these classes, – provided that we can (i) express the defining property of such a class without referring to the concept of order and (ii) define the relation "b before c" in question by means of that class.

An elegant way of doing this, due to W. Sierpiński [1921] consists in taking the class of those initial sections relative to an order of a set a which have a last element. The defining property of such a class D is expressible as follows

Df 2.5 $\mathrm{OrS}(D, a) \;\leftrightarrow\; (x)(y)(x \in D \mathbin{\&} y \in D \rightarrow x \subseteq y \vee y \subseteq x) \mathbin{\&}$
$\mathbin{\&} (x)(x \in D \rightarrow x \subseteq a \mathbin{\&} (Ez)(u)(u = z \leftrightarrow u \in x \mathbin{\&}$
$\mathbin{\&} (y)(y \in D \mathbin{\&} u \in y \rightarrow x \subseteq y))) \mathbin{\&}$
$\mathbin{\&} (z)(z \in a \rightarrow (Ex)(x \in D \mathbin{\&} z \in x \mathbin{\&} (y)(y \in D \mathbin{\&}$
$\mathbin{\&} z \in y \rightarrow x \subseteq y))).$

The relation "b before c" is expressed by the formula

$$(Ex)(x \in D \mathbin{\&} b \in x \mathbin{\&} c \notin x).$$

An immediate consequence of Df 2.5 is

2.6 $\mathrm{OrS}(D, a) \mathbin{\&} K \equiv \{zx \mid x \in D \mathbin{\&} z \in x \mathbin{\&} (y)(y \in D \mathbin{\&}$
$\mathbin{\&} z \in y \rightarrow x \subseteq y)\}. \;\cdot\!\rightarrow\!\cdot\; \mathrm{Crs}(K) \mathbin{\&} a^* \,\overline{\,{}_K\,}\, D$

[1] This was first observed by Hessenberg [1906] and further carried out by Kuratowski [1921].

and therefore also by II, 3.8

2.7 $$\mathrm{OrS}(D, a) \rightarrow \mathrm{Rp}(D).$$

Thus also by the order concept OrS every order of a set is represented by a set. In this respect Or and OrS behave alike. However there is the difference that the definition of $\mathrm{OrS}(D, a)$ cannot be extended like that of $\mathrm{Or}(C, a)$ to the case of a class instead of the set a to be ordered.

The interrelation between the two order concepts, applied to the ordering of a set a is expressed by the provable formulas

2.8 $$\mathrm{Or}(C, a) \;\&\; D \equiv \{x \,\|\, (Ey)(y \in a \;\&\; (z)(z \in x \leftrightarrow \langle z, y\rangle \in C))\} \rightarrow \rightarrow \mathrm{OrS}(D, a) \;\&\; (\langle b, c\rangle \in C \leftrightarrow c \in a \;\&\; (z)(z \in D \;\&\; c \in z \rightarrow b \in z)),$$

2.9 $$\mathrm{OrS}(D, a) \;\&\; C \equiv \{xy \,\|\, y \in a \;\&\; (z)(z \in D \;\&\; y \in z \rightarrow x \in z)\} \rightarrow \rightarrow \mathrm{Or}(C, a) \;\&\; (c \in D \leftrightarrow (Ey)(y \in a \;\&\; (z)(z \in c \leftrightarrow \langle z, y\rangle \in C))).$$

Going on now to the comparison of orders (of classes or sets) we first state the theorem

2.10 $$\mathrm{Or}(C, A) \;\&\; A \,\overline{\overline{{}_K}}\, B \rightarrow \mathrm{Or}(\breve{K} \,|\, C \,|\, K, B).$$

«For any order C of a class A and any one-to-one correspondence K of A with a class B the pairclass $\breve{K} \,|\, C \,|\, K$ is an order of B».

The proof goes by using that we have upon our premise

$$\breve{K} \,|\, C \,|\, K \equiv \{xy \,\|\, (Eu)(Ev)(\langle u, v\rangle \in C \;\&\; u = \breve{K}`x \;\&\; y = K`v)\}$$

and also $\mathrm{Crs}(K)$. Let us call the order of B by $\breve{K}|C|K$ the order *induced* from the order C of A by the one-to-one mapping K; it is that order of B by which $K`a$ is before $K`b$ if and only if a is before b by the order C of A.

For any one-to-one correspondence K between A and B the connection between an order C of A and the order D of B which is induced by K from C has the character of an equivalence relation. Indeed defining

Df 2.6 $$C, A \,\overline{\overline{\overline{{}_K}}}\, D, B \leftrightarrow \mathrm{Or}(C, A) \;\&\; A \,\overline{\overline{{}_K}}\, B \;\&\; D \equiv \breve{K} \,|\, C \,|\, K,$$

we have

$$C, A \mathrel{\underset{\{xy \,|\, x=y\}}{\sqcap}} C, A$$

2.11 $$C, A \mathrel{\underset{K}{\sqcap}} D, B \rightarrow D, B \mathrel{\underset{\breve{K}}{\sqcap}} C, A$$

$$C, A \mathrel{\underset{K}{\sqcap}} D, B \;\&\; D, B \mathrel{\underset{L}{\sqcap}} Q, P \;\rightarrow\; C, A \mathrel{\underset{K|L}{\sqcap}} Q, P.$$

By means of Df 2.6 we now define similarity between orders of sets using that any one-to-one correspondence between sets is represented by a set. The definition is

Df 2.7 $$c \approx d \leftrightarrow (Ex)(c^*, \Delta_1 c^* \mathrel{\underset{x^*}{\sqcap}} d^*, \Delta_1 d^*)$$

«c and d are equal orders».

From this we first get

2.12 $$c \approx d \rightarrow \mathrm{Or}(c) \;\&\; \mathrm{Or}(d)$$

2.13 $$c \approx d \rightarrow \Delta_1 c \sim \Delta_1 d.$$

Further the relation $c \approx d$, as follows from 2.11, has the formal properties of a kind of equality, so that we have

2.14 $$\mathrm{Or}(c) \;\&\; \mathrm{Or}(d) \cdot\!\rightarrow\!\cdot\, c \approx d \leftrightarrow \{x \,|\, c \approx x\} \equiv \{x \,|\, d \approx x\}.$$

The class $\{x \,|\, c \approx x\}$ which for any order c is the class of the orders equal to c can be taken as the *order type* of c. But we are not able to represent order types thus defined by sets.

A simple instance of an ordering class is

$$\{xy \,|\, \mathrm{Od}(y) \;\&\; (x \in y \vee x = y)\}$$

by which the *natural order* of the class of ordinals is characterized, which indeed is the order by "lower" and "higher" or what comes out to the same, the order by the element-relation. This formally results by applying the formula III, 1.5 and III, 1.11. Generally we understand by the natural order of any class of ordinals A the intersection of the above class with $A \times A$, and by the natural order of a set of ordinals a the natural order of a^*, which indeed is represented by an ordering set. In order to have a brief notation we define

Df 2.8 $$\mathrm{no}(a) = a \times a \cap \{xy \,|\, \mathrm{Od}(y) \;\&\; (x \in y \vee x = y)\}.$$

For any set a of ordinals, $\mathrm{no}(a)$ represents the natural order of a

and we have

2.15 $$\mathrm{Or}(\mathrm{no}(a)).$$

We mention still that by the Sierpiński definition of order the natural order of a set a of ordinals is the class

$$\{x \,|\, (Ez)(z \in a \;\&\; x = a \cap z')\}.$$

A natural generalization of the concept of order which comes to be used in many domains of mathematics is that of *partial order*. The definition which results from Df 2.1 by weakening the conditions is

Df 2.9 $$\mathrm{Po}(C, A) \leftrightarrow C \subseteq A \times A \;\&\; (x)(x \in A \to \langle x, x\rangle \in C) \;\& \\ \& \; (x)(y)(\langle x, y\rangle \in C \;\&\; \langle y, x\rangle \in C \to x = y) \;\& \\ \& \; (x)(y)(z)(\langle x, y\rangle \in C \;\&\; \langle y, x\rangle \in C \to \langle x, z\rangle \in C).$$

«C is a partial order of A.»

Quite corresponding is the definition of $\mathrm{Po}(C, a)$; and we also define again

Df 2.10 $$\mathrm{Po}(c, a) \leftrightarrow \mathrm{Po}(c^*, a).$$

All the properties of Or stated by the formulas 2.1–2.3 hold likewise for Po. In particular every partial order of a set is represented by a set. Also the definitions Df 2.3, Df 2.4 apply likewise to the concept Po. These definitions of $\mathrm{Po}(C)$, $\mathrm{Po}(c)$ are used later on. Evidently every order is also a partial order, but not inversely. Like every order of a class A also every partial order can be restricted to any subclass of A. In case that a partial order C restricted to a certain subclass of its domain is a full order, we call this order a *chain* of C. The defining condition of D being a chain of C is

Df 2.11 $$\mathrm{Ch}(D, C) \leftrightarrow \mathrm{Po}(C) \;\&\; D \subseteq C \;\&\; \mathrm{Or}(D).$$

We apply the concept of chain likewise to sets which are partial orders, defining

Df 2.12 $$\mathrm{Ch}(d, c) \leftrightarrow \mathrm{Ch}(d^*, c^*).$$

When S is a subclass of the domain $\Delta_1 C$ of a partial order C such that $\mathrm{Or}(C \cap (S \times S))$ and thus $\mathrm{Ch}(C \cap (S \times S), C)$, we shall say that S determines a chain in C. The necessary and sufficient condition for this is that for any two elements d, e of S we have

$$\langle d, e\rangle \in C \vee \langle e, d\rangle \in C.$$

An essential specialization of the concept of partial order still much more general than that of full order is that of a *lattice*. As well known, the restricting condition of a lattice over a partial order is that of the existence for any two elements of a least upper bound and a greatest lower bound. Formally the definition is

Df 2.13 $\mathrm{La}(C) \leftrightarrow \mathrm{Po}(C)\ \&\ (x)(y)(x \in \Delta_1 C\ \&\ y \in \Delta_1 C \rightarrow (Ez)(\langle x,z\rangle \in C\ \&$
$\&\ \langle y, z\rangle \in C\ \&\ (u)(\langle x, u\rangle \in C\ \&\ \langle y, u\rangle \in C \rightarrow$
$\rightarrow \langle z,u\rangle \in C))\ \&\ (Ez)(\langle z,x\rangle \in C\ \&\ \langle z,y\rangle \in C\ \&$
$\&\ (u)(\langle u, x\rangle \in C\ \&\ \langle u, y\rangle \in C \rightarrow \langle u, z\rangle \in C))),$

and again

Df 2.14 $\mathrm{La}(c) \leftrightarrow \mathrm{La}(c^*).$

V, § 3. Wellorder

In considering wellorder we start from the intuitive concept of wellorder (to which we referred in Chapt III, § 1). According to it an order C of a class A is a wellorder if every non empty subclass of A has a foremost element with respect to C. (This among the various possible characterizations of wellorder is perhaps not the most pregnant from the structural point of view but that one most easy to formulate and very handy for the applications.)

For a direct formalization of this condition of wellorder we should have to use bound class variables, what we are avoiding in our formal system.

However we first observe that in the case where the ordered class is represented by a set a, the condition of wellorder is not weakened by requiring only that every non empty subset of a has a foremost element, what of course is expressible in our formal frame. Indeed the condition on the subsets amounts here to the same as that one on the subclasses, since every subclass of a is represented by a subset of a.

But even for the case that a representation of the ordered class A is not available, the condition that every non empty subset of A has a foremost element can be proved to imply the same for every non empty subclass, but this only upon a strengthening of our formal system, namely by adding a strong form of the axiom of choice and the axiom of infinity. In fact with the aid of these axioms we shall be able to prove that if A is a class ordered by C and there is a subclass of A which is not empty and, by the order C, has no foremost element, then there also exist a subset of A with the same property. (Cf. chapt. VIII, end of § 1.)

Now coming back to the case of wellorder of a set, we formulate two definitions of wellorder, corresponding to the two concepts Or and OrS:

Df 3.1 $\quad \mathrm{Wor}(C, a) \leftrightarrow \mathrm{Or}(C, a) \mathbin{\&} (x)(x \subseteq a \mathbin{\&} x \neq 0 \rightarrow (Eu)(u \in x \mathbin{\&}$
$$\mathbin{\&} (v)(v \in x \rightarrow \langle u, v\rangle \in C)))$$

Df 3.2 $\quad \mathrm{WorS}(C, a) \leftrightarrow \mathrm{OrS}(C, a) \mathbin{\&} (x)(x^* \subseteq C \mathbin{\&} x \neq 0 \rightarrow$
$$\rightarrow \bigcap_z (z \in x) \in x).$$

These two concepts are related to one another in quite the corresponding way as Or and OrS are. That is to say: if in the formulas 2.8, 2.9 we set Wor in the places of Or and WorS in the places of OrS, the resulting formulas are again provable.

We shall however not have to make much use of either of them. In fact we are to see now that the consideration of wellorders of sets can fully be reduced to that of numerations of sets, as defined in IV, § 3. This will be done by setting up explicitly a function by which the class of numerations of a set a is mapped in a one-to-one way onto the class of wellorders of a.

For this purpose we take the definition of wellorder by Wor. First we note that by 2.3 we have

3.1 $$\mathrm{Wor}(C, a) \rightarrow \mathrm{Rp}(C)$$

and we define parallel to Df 2.2, Df 2.3

Df 3.3 $$\mathrm{Wor}(c, a) \leftrightarrow \mathrm{Wor}(c^*, a)$$ [1].

Df 3.4 $$\mathrm{Wor}(c) \leftrightarrow \mathrm{Wor}(c, \Delta_1 c)$$

Further using 2.15 and the formula

$$(x)(x \in a \rightarrow \mathrm{Od}(x))\ \&\ a \neq 0 \ \rightarrow\ (Ez)(z \in a\ \&\ (x)(x \in a \rightarrow z = x \vee z \in x))$$

which directly follows from the principle of the least ordinal number III, 1.12, we get the formula

3.2 $$(x)(x \in a \rightarrow \mathrm{Od}(x)) \ \rightarrow\ \mathrm{Wor}(\mathrm{no}(a), a)$$

which expresses that the natural order of a set of ordinals is a wellorder. From this in particular we infer

3.3 $$\mathrm{Od}(n) \rightarrow \mathrm{Wor}(\mathrm{no}(n), n)$$

«The natural order of an ordinal is a wellorder».

On the other hand, by means of Df 3.1 together with 2.10 we get

3.4a $$\mathrm{Wor}(C, a)\ \&\ a^* \,\overline{\overline{\scriptstyle K}}\, b^* \ \rightarrow\ \mathrm{Wor}(\breve{K}\,|\,C\,|\,K, b).$$

which also entails

3.4b $$\mathrm{Wor}(a)\ \&\ a \approx b \rightarrow \mathrm{Wor}(b).$$

Applying this formula and the definition of numeration IV, Df 3.1 we obtain

3.5 $$\mathrm{Num}(s, a)\ \&\ \Delta_1 s = n \ \rightarrow\ \mathrm{Wor}(\breve{s}\,|\,\mathrm{no}(n)\,|\,s, a).$$

Thus the order of a set a which is induced by a numeration of a with the domain n from the natural order of n is a wellorder of a, or in other words the function

[1] $$\{xy \,\|\, \mathrm{Num}(x, a)\ \&\ y = \breve{x}\,|\,\mathrm{no}(\Delta_1 x)\,|\,x\}$$

— let us denote it by $\mathfrak{R}(a)$ — is a mapping of the class of numerations of a into the class of wellorders of a.

Next we are to show that this mapping is onto the class of wellorders of a, i.e. that every wellorder of a set a is induced by a numeration of a from the natural order of its domain n.

---

[1]) Our convention of the footnote in § 2 concerning order will be applied also to wellorders.

Let d be a wellorder of a so that we have Wor(d, a). Then by Df 3.1 we have

$$t \subset a \rightarrow (Eu)(u \in a^{-}t \;\&\; (v)(v \in a^{-}t \rightarrow \langle u, v\rangle \in d)).$$

If this is briefly indicated by

$$t \subset a \rightarrow (Eu)\mathfrak{A}(u, a, t)$$

we also have

$$\mathfrak{A}(c, a, t) \;\&\; \mathfrak{A}(e, a, t) \;\rightarrow\; c = e.$$

Therefore the class $\{xy \,\|\, x \subset a \;\&\; y = \iota_u \mathfrak{A}(u, a, x)\}$ is a function which satisfies the conditions of the theorem of generated numeration IV, 3.1. So by this theorem there exists a numeration of a generated by that function, i.e. a one-to-one correspondence s between an ordinal n and the set a such that

$$k \in n \;\&\; \Delta_2\mathrm{sg}(s, k) = t \;\rightarrow\; s`k = \iota_u \mathfrak{A}(u, a, t)$$

and therefore also

[2] $$\Delta_2\mathrm{sg}(s, k) = t \;\rightarrow\; (v)(v \in a^{-}t \rightarrow \langle s`k, v\rangle \in d).$$

Now by this one-to-one correspondence s the wellorder d of a is induced from the natural order of n or formally expressed

$$d = \breve{s} \mid \mathrm{no}(n) \mid s.$$

For proving this, it is sufficient, in virtue of Crs(s), $\Delta_2 s = a$, Wor(d, a), to show that

$$k \in l \;\&\; l \in n \;\rightarrow\; \langle s`k, s`l\rangle \in d.$$

But upon the premise $k \in l \;\&\; l \in n$ we have

$$\Delta_2\mathrm{sg}(s, k) = t \;\rightarrow\; s`l \in a^{-}t$$

and therefore by [2] $\langle s`k, s`l\rangle \in d$. Thus we have

3.6 $$\mathrm{Wor}(d, a) \;\rightarrow\; (Ez)(\mathrm{Num}(z, a) \;\&\; d = \breve{z} \mid \mathrm{no}(\Delta_1 z) \mid z),$$

i.e. every wellorder of a set a is a value of the function $\mathfrak{K}(a)$.

Still it remains to show that the function $\mathfrak{K}(a)$ is a one-to-one correspondence. This comes out to prove

3.7 $$\mathrm{Num}(s, a) \;\&\; \mathrm{Num}(t, a) \;\&\; \Delta_1 s = n \;\&\; \Delta_1 t = m \cdot\!\rightarrow\!\cdot$$
$$\cdot\!\rightarrow\!\cdot\; \breve{s} \mid \mathrm{no}(n) \mid s = \breve{t} \mid \mathrm{no}(m) \mid t \rightarrow s = t.$$

The proof goes as follows. Our premises entail that

$$t \mid \breve{s} \mid \mathrm{no}(n) \mid s \mid \breve{t} = \mathrm{no}(m).$$

Thus since $t \mid \breve{s} = \breve{(s \mid \breve{t})}$, no($m$) is induced from no($n$) by the one-to-one correspondence $s \mid \breve{t}$. If this one-to-one correspondence is an identical mapping, then it follows that $s = t$. So it is sufficient to show that the natural order of an ordinal n cannot pass into the natural order of an ordinal m by a nonidentical one-to-one correspondence L. But in the other case there would be by the principle of least ordinal number, among the elements of n a lowest one k such that $L\text{`}k$ is higher then $k$ and also that $\breve{L}\text{`}k$ is higher than k; therefore if $\breve{L}\text{`}k = l$ we should have

$$L\text{`}l = k, \quad \breve{L}\text{`}k = l$$

$$k \in L\text{`}k, \quad k \in \breve{L}\text{`}k$$

therefore

$$L\text{`}l \in L\text{`}k, \quad k \in l$$

and so the order of m induced by L from no(n) would not be the natural order.

Thus we find that indeed $\mathfrak{K}(a)$ is a one-to-one mapping of the class of numerations of a onto the class of wellorders of a; formally

3.8 $$\{x \mid \mathrm{Num}(x, a)\} \underset{\mathfrak{K}(a)}{\overline{\quad}} \{x \mid \mathrm{Wor}(x, a)\}$$

with $\mathfrak{K}(a)$ being the class term [1].

Our result 3.8 includes the statement that the order type of any wellorder d of a set a is the same as that one of the natural order of the ordinal which is the domain of the numeration, to which d is assigned by $\mathfrak{K}(a)$, or formally

3.9 $$\mathrm{Wor}(d, a) \rightarrow d \approx \mathrm{no}(\Delta_1(\breve{\mathfrak{K}(a)}\text{`}d)).$$

By this way the theory of wellorders and wellorder types of sets is fully reduced to the theory of ordinals. In particular it results that the question if every set can be wellordered is equivalent to that if every set can be numerated.

We add here still a remark concerning the numerations of subsets of ordinals. Let t be a subset of an ordinal n; then to the natural order of t, which indeed is a wellorder, corresponds a numeration s of t such that

$$k \in \Delta_1 s \;\&\; l \in \Delta_1 s \;\&\; k \in l \rightarrow s{}^{\iota}k \in s{}^{\iota}l.$$

Now the ordinal $\Delta_1 s$ is not higher than n. For otherwise there would be in $\Delta_1 s$ a least ordinal k such that $s{}^{\iota}k \in k$. But then we should have $s{}^{\iota}(s{}^{\iota}k) \in s{}^{\iota}k$ what gives a contradiction. Thus we have that for every subset of an ordinal n there is a one-to-one correspondence with an ordinal not higher than n, which besides preserves the natural order, or formally

3.10 $$\mathrm{Od}(n) \;\&\; t \subseteq n \rightarrow (Ex)(\mathrm{Num}(x, t) \;\&\; \Delta_1 x \subseteq n \;\& \\ \& \;\mathrm{no}(t) = \breve{x} \mid \mathrm{no}(\Delta_1 x) \mid x).$$

3.10 includes the statement that for every set of ordinals t there is a numeration by the natural order. Indeed we have $t \subseteq (\sum t)'$ and $(\sum t)'$ is an ordinal.

This statement together with IV, 3.3 yields the following alternative on any class C of ordinals: Either C is represented by a set c and then there is a numeration of c by the natural order, or else for every ordinal k there is a higher ordinal in C and then there is a one-to-one correspondence between $\{x \mid \mathrm{Od}(x)\}$ and C which preserves the natural order.

CHAPTER VI

THE COMPLETING AXIOMS

VI, § 1. The potency axiom

In our preceding discussions of general set theory several times appeared for certain reasonings the need of a complementation of our axiomatical basis. The axioms here in question are the potency axiom, the axiom of choice and the axiom of infinity. Now going on to the complementation of our axiomatic basis we have to consider the contributions afforded by these axioms. We begin with stating and discussing the axioms themselves, each of which can be formulated in various different forms.

Let us consider the *potency axiom*. We take it in a form similar to Zermelo's Potenzmengenaxiom, but introducing the unary function symbol $\pi(a)$, with the axiom

A 4 $$c \in \pi(a) \leftrightarrow c \subseteq a$$

«The elements of $\pi(a)$ are the subsets of a.»

By this axiom it is postulated that the class $\{x \mid x \subseteq a\}$ is represented by a set. In fact we have by it

1.1 $$\pi(a)^* \equiv \{x \mid x \subseteq a\}.$$

That A 4 is called potency axiom is motived by its enabling us to introduce $a^{\underline{b}}$ as a set representing $(a^*)^{\underline{b}}$. In fact by A 4 we can prove that the class $(a^*)^{\underline{b}}$, i.e. by II, Df 4.9 the class of those functional sets with the domain b whose converse domain is a subset of a^*, is represented by a set. For, each functional set of the said kind is a subset of $b \times a$, hence $(a^*)^{\underline{b}}$ is a subclass of $\pi(b \times a)$ and so it is represented by $\pi(b \times a) \cap (a^*)^{\underline{b}}$. We therefore can define

Df 1.1 $$a^{\underline{b}} = \iota_x(x^* \equiv (a^*)^{\underline{b}})$$

and we have

1.2 $$(a^{\underline{b}})^* \equiv (a^*)^{\underline{b}}.$$

There would also be the possibility of taking 1.2 as axiom instead of A 4. In fact from 1.2 we can infer that for any set a the class $\{x \mid x \subseteq a\}$ is represented. Indeed this follows by the theorem of replacement II, 3.8, since there is an obvious one-to-one correspondence between that class and the class represented by $0''\underline{a}$, i.e. the class of Belegungen of a with values 0, 0′. This one-to-one correspondence is

$$\{xy \mid x \subseteq a \ \& \ y^* \equiv \{uv \mid u \in x \ \& \ v = 0 \cdot \mathbf{v} \cdot u \in a^- x \ \& \ v = 0'\}\}.$$

At the same time it results that

1.3 $$\pi(a) \sim 0''\underline{a}.$$

From the existence of the Potenzmenge $\pi(a)$ for an arbitrary set a there further follows the existence of the Cantor *Produktmenge* which is the generalization of the crossproduct $c \times d$ to an arbitrary number of factors. Indeed let the factors be given by the values of a term $\mathfrak{t}(\mathfrak{a})$ for $\mathfrak{a}$ ranging over the elements of a set m. Then their product as a class is given by

$$\{\mathfrak{y} \mid \mathrm{Ft}(\mathfrak{y}) \ \& \ \Delta_1 \mathfrak{y} = m \ \& \ (\mathfrak{x})(\mathfrak{x} \in m \to \mathfrak{y}\text{‘}\mathfrak{x} \in \mathfrak{t}(\mathfrak{x}))\}.$$

Now the elements of this class are subsets of $m \times \sum_{\mathfrak{x}}(m, \mathfrak{t}(\mathfrak{x}))$, and so the class itself is a subclass of $\pi(m \times \sum_{\mathfrak{x}}(m, \mathfrak{t}(\mathfrak{x})))$ and therefore by II, 2.1 represented by a set; hence we can define the Produktmenge by

Df 1.2 $$\prod_{\mathfrak{x}}(m, \mathfrak{t}(\mathfrak{x})) = \iota_{\mathfrak{z}}(\mathfrak{z}^* \equiv \{\mathfrak{y} \mid \mathrm{Ft}(\mathfrak{y}) \ \& \ \Delta_1 \mathfrak{y} = m \ \& \ \& \ (\mathfrak{x})\ (\mathfrak{x} \in m \to \mathfrak{y}\text{‘}\mathfrak{x} \in \mathfrak{t}(\mathfrak{x}))\}$$

and we have

1.4 $$a \in \prod_{\mathfrak{x}}(m, \mathfrak{t}(\mathfrak{x})) \leftrightarrow \mathrm{Ft}(a) \ \& \ \Delta_1 a = m \ \& \ (\mathfrak{x})(\mathfrak{x} \in m \to a\text{‘}\mathfrak{x} \in \mathfrak{t}(\mathfrak{x})).$$

Thus the symbol $\prod_{\mathfrak{x}}(m, \mathfrak{t}(\mathfrak{x}))$ figures in full analogy to the symbol $\sum_{\mathfrak{x}}(m, \mathfrak{t}(\mathfrak{x}))$. This analogy also shows that we could formalize the concept of the Produktmenge with a class variable F, thus writing

$\prod(m, F)$ in analogy to II, Df 1.4; instead of the schema 1.4 we then would have the formula

$$a \in \prod(m, F) \leftrightarrow \mathrm{Ft}(a) \,\&\, \Delta_1 a = m \,\&\, (x)(x \in m \rightarrow a^{\iota}x \in F^{\iota}x)$$

which indeed entails every application of 1.4 in virtue of II, 1.15.

The concept of the Produktmenge can be regarded as a generalization of that of $a^{\underline{b}}$. In fact we have

1.5 $$a^{\underline{b}} = \prod_x(b, \iota_z(x = x \,\&\, z = a)).$$

From 1.5 together with the above statement on the derivability of A 4 from 1.2 it follows that A 4 can also be derived from 1.4, so that the assumption that the Produktmenge is represented could be taken instead of the potency axiom.

A main effect of introducing the potency axiom is that for the mathematical applications of our axiomatic system the handling with classes can almost everywhere be avoided, since all occurring classes are represented by sets.

Let us at once note some immediate applications of A 4 where this axiom enables us to complete some former statements.

The formulas II, 4.13–II, 4.16 in combination with 1.2 and II, 3.10 yield

1.6 $$a \sim b \,\&\, c \sim d \rightarrow a^{\underline{c}} \sim b^{\underline{d}}$$

1.7 $$a^{\underline{c}} \times b^{\underline{c}} \sim (a \times b)^{\underline{c}}$$

1.8 $$b \cap c = 0 \rightarrow a^{\underline{b}} \times a^{\underline{c}} \sim a^{\underline{b \cup c}}$$

1.9 $$(a^{\underline{b}})^{\underline{c}} \sim a^{\underline{b \times c}}.$$

From the formulas V, 1.9 together with the obvious formula

$$\{x \mid (Ez)(z \in a \,\&\, x = [z]\} \subseteq \pi(a)^*$$

we get

$$(Eu)(u \sim a \,\&\, u \subseteq \pi(a))$$

i.e. by Df V, 1.1

[1] $$a \precsim \pi(a).$$

On the other hand V, 1.8 directly entails

$$\overline{(Ex)}((y)(y \subseteq a \rightarrow y \in x) \mathbin{\&} a \sim x)$$

and thus also

[2] $$\overline{a \sim \pi(a)}.$$

[1] and [2] together give

1.10 $$a \prec \pi(a).$$

In this formula is included the statement that to every set there exists a set of higher power.

VI, § 2. The axiom of choice

For the introduction of the axiom of choice we first discuss various forms of stating the choice principle.

We start from that form of its statement which is related to the Produktmenge. Here it expresses the intuitively convincing assertion that a Produktmenge $\prod\limits_{\mathfrak{x}}(m, \mathfrak{t}(\mathfrak{x}))$ is empty only if at least one factor $\mathfrak{t}(\mathfrak{x})$ is zero. The formal statement is by the schema

2.1 $$(\mathfrak{x})(\mathfrak{x} \in m \rightarrow \mathfrak{t}(\mathfrak{x}) \neq 0) \rightarrow \prod_{\mathfrak{x}}(m, \mathfrak{t}(\mathfrak{x})) \neq 0.$$

In virtue of the defining property of the Produktmenge Df 1.2 this comes out to

2.2 $$(\mathfrak{x})(\mathfrak{x} \in m \rightarrow \mathfrak{t}(\mathfrak{x}) \neq 0) \rightarrow$$
$$\rightarrow (E\mathfrak{y})(\mathrm{Ft}(\mathfrak{y}) \mathbin{\&} \Delta_1\mathfrak{y} = m \mathbin{\&} (\mathfrak{x})(\mathfrak{x} \in m \rightarrow \mathfrak{y}{}^{\prime}\mathfrak{x} \in \mathfrak{t}(\mathfrak{x}))).$$

Instead of this schema we can as well take the formula

2.3 $$\mathrm{Ft}(f) \mathbin{\&} (x)(x \in m \rightarrow f{}^{\prime}x \neq 0) \rightarrow (Ey)(\mathrm{Ft}(y) \mathbin{\&} \Delta_1 y = m \mathbin{\&}$$
$$\mathbin{\&} (x)(x \in m \rightarrow y{}^{\prime}x \in f{}^{\prime}x)).$$

«For every function f whose values for the elements of m are non empty sets there exists a function assigning to every element x of m an element of $f{}^{\prime}x$».

Indeed on one hand $\mathfrak{t}(\mathfrak{x})$ can be taken to be $f`x$; on the other hand 2.2 results from 2.3 since for every term $\mathfrak{t}$ the class

$$\{\mathfrak{x}\mathfrak{y} \mid \mathfrak{x} \in m \ \& \ \mathfrak{y} = \mathfrak{t}(\mathfrak{x})\}$$

is by II, 3.10 represented by a functional set f with the domain m.

An equivalent formulation of 2.3 is obtained by substituting g for f and $\Delta_1 g$ for m:

2.4 $\mathrm{Ft}(g) \ \& \ (x)(z)(\langle x, z\rangle \in g \rightarrow z \neq 0) \cdot\!\!\rightarrow\!\cdot$

$\cdot\!\rightarrow\!\cdot (Ey)(\mathrm{Ft}(y) \ \& \ \Delta_1 y = \Delta_1 g \ \& \ (x)(u)(\langle x, u\rangle \in g \rightarrow y`x \in u)).$

For going back from 2.4 to 2.3 we have to substitute for g the term $f \cap (m \times V)$ i.e. to take for g the function f restricted to the domain m.

Applying 2.2 to the case that $\mathfrak{t}(\mathfrak{x})$ is simply $\mathfrak{x}$ we get

2.5 $(x)(x \in m \rightarrow x \neq 0) \rightarrow (Ey)(\mathrm{Ft}(y) \ \& \ \Delta_1 y = m \ \&$

$\& \ (x)(x \in m \rightarrow y`x \in x)).$

«If m is a set of non empty sets, there exists a function assigning to each element of m one of its elements».

From 2.5 Russell's and Zermelo's multiplicative axiom can be derived. This goes by specializing 2.5 to the case where the elements of m are mutually exclusive sets. Here then the converse domain of the function y stated to exist by 2.5, is a set which has just one element in common with every element of m. Formalizing this argument by the predicate calculus together with the equality axioms and using the theorem of replacement—there is no trick in the derivation—we come to prove the formula

2.6 $(x)(y)(x \in m \ \& \ y \in m \rightarrow x \neq 0 \ \& \ (x = y \lor x \cap y = 0) \cdot\!\!\rightarrow\!\cdot$

$\cdot\!\rightarrow\!\cdot (Ev)(x)(x \in m \rightarrow (Eu)(u \in x \cap v \ \& \ (z)(z \in x \cap v \rightarrow z = u))).$

«If m is a set of mutually exclusive non empty sets then there exists a set v which has with every element of m just one element in common, i.e. there exists for such a set m an Auswahlmenge».

An other form of a choice principle often used is the following

2.7 $$\mathrm{Ps}(a) \rightarrow (Ey)(y \subseteq a \mathbin{\&} \Delta_1 y = \Delta_1 a \mathbin{\&} \mathrm{Ft}(y)).$$

«For every set of pairs a there exists a subset which is a function with the same domain as a».

The formula 2.7 can be derived from 2.6 as follows. Let $\mathfrak{C}$ be the class term

[1] $$\{uv \mid u \in \Delta_1 a \mathbin{\&} v = a \cap \{z \mid (Ey)(\langle u, y\rangle = z)\}\},$$

then we have

$$\mathrm{Ps}(a) \rightarrow \mathrm{Ft}(\mathfrak{C}) \mathbin{\&} \Delta_1 \mathfrak{C} \equiv \Delta_1 a^*$$

and also by II, 3.8

$$\mathrm{Ps}(a) \rightarrow \mathrm{Rp}(\Delta_2 \mathfrak{C}), \qquad \text{i.e.}$$

[2] $$\mathrm{Ps}(a) \rightarrow (Ew)(w^* \equiv \Delta_2 \mathfrak{C}).$$

Further we have

$$\mathrm{Ps}(a) \mathbin{\&} m^* \equiv \Delta_2 \mathfrak{C} \rightarrow (x)(y)(x \in m \mathbin{\&} y \in m \rightarrow x \neq 0 \mathbin{\&} \\ \mathbin{\&} (x = y \vee x \cap y = 0))$$

and therefore by 2.6

$$\mathrm{Ps}(a) \mathbin{\&} m^* \equiv \Delta_2 \mathfrak{C} \rightarrow (Ev)(x)(x \in m \rightarrow (Eu)(z)(z = u \leftrightarrow z \in x \cap v))$$

what together with [2] yields

$$\mathrm{Ps}(a) \rightarrow (Ev)(x)(x \in \Delta_2 \mathfrak{C} \rightarrow (Eu)(z)(z = u \leftrightarrow z \in v \cap x)).$$

From this using the expression [1] of $\mathfrak{C}$ we get

[3] $$\mathrm{Ps}(a) \rightarrow (Ev)((u)(u \in \Delta_1 a \rightarrow (Ex)(z)(z = x \leftrightarrow \\ \leftrightarrow z \in v \cap (a \cap \{w \mid (Ey)(\langle u, y\rangle = w)\}))).$$

«For every set of pairs a there exists a set v such that for every element u of the domain of a, v has just one element in common with the class of those pairs out of a whose first element is u».

Let us indicate the formula [3] by $\mathrm{Ps}(a) \rightarrow (Ev)(\mathfrak{B}(a, v)$. Now there is no difficulty in proving

$$\mathrm{Ps}(a) \,\&\, \mathfrak{B}(a, f) \rightarrow f \cap a \subseteq a \,\&\, \Delta_1(f \cap a) = \Delta_1 a \,\&\, \mathrm{Ft}(f \cap a)$$

and this together with [3] yields 2.7.

The formula 2.7 can in particular be applied to the case where a is the converse $\breve{f}$ of a function f so that $\Delta_1 a = \Delta_2 f$. Hence we come to the formula

2.8 $$\mathrm{Ft}(f) \rightarrow (Ey)(y \subseteq \breve{f} \,\&\, \Delta_1 y = \Delta_2 f \,\&\, \mathrm{Ft}(y)).$$

The content of this formula can briefly be resumed by the statement that for every functional set there exists an inverse functional set. By the way, a functional set which is inverse to a functional set must be a one-to-one correspondence, what follows from $\mathrm{Ft}(f) \,\&\, g \subseteq f \rightarrow \mathrm{Ft}(g)$; so in 2.8 the conjunction member $\mathrm{Ft}(y)$ could be replaced by $\mathrm{Crs}(y)$.

The statement 2.8, though obtained as a consequence of specializations of 2.2, is nevertheless no more special than 2.2. In fact we are to derive the schema 2.2 from 2.8 in the following way.

Let $\mathfrak{C}$ be the class

$$\{\mathfrak{x} \mid (E\mathfrak{u})(E\mathfrak{v})(\mathfrak{u} \in m \,\&\, \mathfrak{v} \in \mathfrak{t}(\mathfrak{u}) \,\&\, \mathfrak{x} = \langle\langle \mathfrak{u}, \mathfrak{v}\rangle, \mathfrak{u}\rangle)\}$$

then we obviously have $\mathrm{Ft}(\mathfrak{C})$ and $(\mathfrak{x})(\mathfrak{x} \in m \rightarrow \mathfrak{t}(\mathfrak{x}) \neq 0) \rightarrow \Delta_2 \mathfrak{C} \equiv m^*$. Moreover

$$\mathfrak{C} \subseteq ((m \times \sum_{\mathfrak{x}}(m, \mathfrak{t}(\mathfrak{x})) \times m)^*$$

so that, by II, 2.1, $\mathrm{Rp}(\mathfrak{C})$. Therefore we get

$$(\mathfrak{x})(\mathfrak{x}) \in m \rightarrow \mathfrak{t}(\mathfrak{x}) \neq 0) \rightarrow (E\mathfrak{z})(\mathrm{Ft}(\mathfrak{z}) \,\&\, \mathfrak{z}^* \equiv \mathfrak{C} \,\&\, \Delta_2 \mathfrak{z} = m).$$

Now 2.8 gives

[1] $$(\mathfrak{x})(\mathfrak{x} \in m \rightarrow \mathfrak{t}(\mathfrak{x}) \neq 0) \rightarrow (E\mathfrak{y})(\mathrm{Ft}(\mathfrak{y}) \,\&\, \mathfrak{y}^* \subseteq \breve{\mathfrak{C}} \,\&\, \Delta_1 \mathfrak{y} = m).$$

On the other hand from the expression of $\mathfrak{C}$ we draw

$$\mathrm{Ft}(g) \,\&\, g^* \subseteq \breve{\mathfrak{C}} \rightarrow (\mathfrak{x})(\mathfrak{x} \in g \rightarrow (E\mathfrak{u})(E\mathfrak{v})(\mathfrak{x} = \langle \mathfrak{u}, \langle \mathfrak{u}, \mathfrak{v}\rangle\rangle \,\&\, \mathfrak{u} \in m \,\&$$
$$\&\, \mathfrak{v} \in \mathfrak{t}(\mathfrak{u}) \,\&\, \mathfrak{v} = \iota_{\mathfrak{z}}(\langle \mathfrak{u}, \mathfrak{z}\rangle = g'\mathfrak{u}),$$

and this further gives

[2] $\mathrm{Ft}(g) \,\&\, g^* \subseteq \breve{\mathfrak{C}} \,\&\, \Delta_1 g = m \cdot\!\rightarrow\!\cdot$
$$\cdot\!\rightarrow\!\cdot \mathrm{Ft}(\Delta_2 g) \,\&\, \Delta_1(\Delta_2 g) = m \,\&\, (\mathfrak{u})(\mathfrak{u} \in m \rightarrow (\Delta_2 g)`\mathfrak{u} \in \mathfrak{t}(\mathfrak{u})).$$

From [1] and [2] together we obtain

$$(\mathfrak{x})(\mathfrak{x} \in m \rightarrow \mathfrak{t}(\mathfrak{x}) \neq 0) \rightarrow (E\mathfrak{y})(\mathrm{Ft}(\mathfrak{y}) \,\&\, \Delta_1 \mathfrak{y} = m \,\&$$
$$\&\, (\mathfrak{x})(\mathfrak{x} \in m \rightarrow \mathfrak{y}`\mathfrak{x} \in \mathfrak{t}(\mathfrak{x}))),$$

which is the same as 2.2.

So all the different considered statements of the choice principle are equivalent, even without using A 4.

In particular it results that 2.7 and 2.8 can derived from one another. This is just the equivalence asserted in V, § 1 between the formulas (E) and (D). From (D), as we found, it can be proved that the two different possible definitions of "at most equal power" come out to the same (cf. V, § 1, B). Also the formula (C) in V, § 1, which is a consequence of (D) is now available; we write it in the equivalent form

2.9 $$\mathrm{Ft}(f) \rightarrow \Delta_2 f \preceq \Delta_1 f.$$

For the statement of the axiom of choice we now have – so to speak – an embarras de richesse. But this is no real difficulty, since from any one of the statements taken as axiom we easily pass to the other forms, which we thus have available as theorems. It seems suitable to adopt the formula 2.7 as our axiom of choice:

A 5 $$\mathrm{Ps}(a) \rightarrow (Ey)(y \subseteq a \,\&\, \Delta_1 y = \Delta_1 a \,\&\, \mathrm{Ft}(y))\ ^{1)},$$

which is especially handy for the derivations. Note in this respect the direct passage from A 5 to 2.5 by taking, upon the premise $(x)(x \in m \rightarrow x \neq 0)$, the set of pairs a to be the set representing the class $\{xy \mid x \in m \,\&\, y \in x\}$.

---

1) This form of the axiom of choice has been used by some authors, for instance by R. Baer [1929].

VI, § 3. THE NUMERATION THEOREM. FIRST CONCEPTS OF CARDINAL ARITHMETIC

As well known, a main application of the axiom of choice is for the wellorder theorem, from which the general comparability of sets with respect to power can be inferred and which therefore constitutes the basis for cardinal arithmetic. In our system by the stated connection between wellorder and numeration the proof of the wellorder theorem is reduced to that of the numeration theorem.

This theorem results as an immediate application of the axiom of choice in the form 2.2. In fact we have already the theorem of generated numeration IV, 3.1 and so, as observed on p. 111, the only thing still needed is to prove that for any set c there exists a function G which assigns to every proper subset p of c an element of $c^{-}p$, (cf. condition [7] on p. 111). This however follows from 2.2 by taking for m: $\pi(c)^{-}[c]$ and for $\mathfrak{t}(a)$: $c^{-}a$. Indeed in this way we get

$$(Ey)(\mathrm{Ft}(y) \ \& \ \Delta_1 y = \pi(c)^{-}[c] \ \& \ (x)(x \in \pi(c)^{-}[c] \rightarrow y^{\iota}x \in c^{-}x)).$$

It is to be observed that we have to use here essentially the axiom A 4. Thus the numeration theorem, i.e. the statement that for every set there exists a numeration follows from the theorem of generated numeration in connection with A 5 and A 4. The formal statement can be given, on account of the definition of numeration IV, Df 3.1 and the concepts involved in it, by the formula

3.1 $$(x)(Ez)(\mathrm{Od}(z) \ \& \ z \sim x).$$

This formula together with III, 1.9 and Df V, 1.1 as also V, 1.2 gives the formula (cf. V, § 1, A)

3.2 $$(x)(y)(x \preceq y \vee y \preceq x)$$

which expresses the general comparability of sets with respect to power.

A further consequence we can draw from 3.1 is that every set can be wellordered

3.3 $$(x)(Ez)(\mathrm{Wor}(z, x)).$$

This in fact results directly from the theorem V, 3.5. On the other hand we have that from every wellorder of a set c we obtain a numeration of it.

A particular consequence of 3.3 is

3.4 $$(x)(Ez)(\mathrm{Or}(z, x)).$$

Of this theorem, though it is weaker than the wellorder theorem, no proof without the axiom of choice is known.

Note also that the axiom of choice can be almost directly inferred from the numeration theorem. For this we apply 3.1 to the converse domain of the set of pairs a occurring in A 5. Let s be a numeration of $\Delta_2 a$, then a subset of a which is a function with the domain $\Delta_1 a$ is obtained by dropping from a all those pairs $\langle b, c\rangle$ to which there is in a a pair $\langle b, d\rangle$ such that d is assigned by s to a lower ordinal than c is assigned to.

In a near connection with the application made of 3.1 to the statements 3.2 and 3.3 is that one to the concept of a *cardinal number*. Following the method of Frege we should have to define the cardinal number of a set a as the class of the sets b such that $a \sim b$; but by this way cardinal numbers in our system would not be sets. Now this difficulty by 3.1 can easily be overcome. Namely according to this theorem for every set a the class $\{x \mid x \sim a\}$ contains at least one ordinal. Further among the ordinals belonging to it there is, by the principle of least ordinal number, a lowest one $\mu_x(x \sim a)$, and by this ordinal the class $\{x \mid x \sim a\}$ is uniquely determined. So we can take this ordinal as the cardinal number of a, which we denote by $\aleph(a)$. The formal definition is

Df 3.1 $$\aleph(a) = \mu_x(x \sim a).$$

From this we immediately get by III, Df 1.5, III, Df 1.6 and 3.1

3.5 $$\mathrm{Od}(\aleph(a)), \quad a \sim \aleph(a)$$

3.6 $$a \sim b \leftrightarrow \aleph(a) = \aleph(b)$$

and hence also

3.7 $$\aleph(\aleph(a)) = \aleph(a).$$

Further with application of V, Df 1.1, V, Df 1.2 and V, 1.2, and the formula 3.6 we get

$$\aleph(a) \subseteq \aleph(b) \rightarrow a \preceq b, \quad \aleph(a) \subset \aleph(b) \rightarrow a \prec b,$$

and from this, using V, 1.6 and III, 1.9

$$a \prec b \rightarrow \aleph(a) \subset \aleph(b)$$

so that we have

3.8 $$a \prec b \leftrightarrow \aleph(a) \subset \aleph(b)$$
$$\leftrightarrow \aleph(a) \in \aleph(b).$$

By 3.6 and 3.8 the relations of equal power and of lower power between sets are reduced to the identity and the $\in$-relation between their cardinal numbers-briefly their cardinals.

We note that, for a finite set a, $\aleph(a)$ is the same as mlt(a). Indeed from Fin(a) $\rightarrow$ Fin($\aleph(a)$), together with 3.5 and III, 5.5, III, 5.7 we get

3.9 $$\text{Fin}(a) \rightarrow \aleph(a) = \text{mlt}(a).$$

Considering the ordinals with respect to their power we are led to the concept of the Cantor *Zahlenklasse*. By the relation of equal power the class $\{x \mid \text{Od}(x)\}$ is divided in mutually exclusive subclasses which we call Zahlenklassen. Thus we define

Df 3.2 $$\text{Nc}(A) \leftrightarrow (Ex)(\text{Od}(x) \;\&\; (y)(y \in A \leftrightarrow \text{Od}(y) \;\&\; y \sim x)).$$

Hence the Zahlenklasse to which an ordinal n belongs is the class of all ordinals which have the same cardinal as n. This cardinal is itself an ordinal belonging to this Zahlenklasse and among these the lowest one. It is called the *initial number* of this Zahlenklasse. The ordinals which figure as initial numbers of Zahlenklassen are thus the same as those which occur as cardinal numbers of sets. Among the ordinals the cardinal numbers of sets are distinguished by the property, that they are not of equal power with a lower ordinal. Hence we can define

Df 3.3 $\quad \mathrm{Cd}(m) \leftrightarrow \mathrm{Od}(m) \;\&\; (x)(x \sim m \rightarrow x \notin m)$
«m is a cardinal number»,

and we have

3.10 $\begin{cases} \mathrm{Cd}(m) \leftrightarrow \aleph(m) = m, \\ \mathrm{Cd}(\aleph(a)). \end{cases}$

With using A 4 we are able to prove that the class of numerations of a set a and likewise the class of numerations of subsets of a is represented by a set. By V, 3.9 there is a one-to-one correspondence between the numerations of a subset t of a and the wellorders of t. Hence by II, 3.8 it is sufficient to prove that the class of wellorders of subsets of a is represented by a set. But this follows from the fact that every wellorder of a subset of a is $\subseteq a \times a$ and therefore the class of all these wellorders is a subclass of $\pi(a \times a)$. Thus we have

3.11 $\quad \mathrm{Rp}(\{ x \mid (Ey)(y \subseteq a \;\&\; \mathrm{Num}(x, y))\}).$

As a corollary of 3.11 we infer that every Zahlenklasse is represented by a set:

3.12 $\quad \mathrm{Nc}(A) \rightarrow \mathrm{Rp}(A).$

Indeed if a is the initial number of a Zahlenklasse A, then the elements of A are the ordinals c such that $c \sim a$. Such a one-to-one correspondence is a numeration of a with the domain c. Thus every element of A occurs as the domain of a numeration of a. Hence A is the converse domain of the function

$$\{xy \mid \mathrm{Num}(x, a) \;\&\; y = \Delta_1 x\}$$

whose domain by 3.11 is represented by a set, therefore A is also represented.

Still as an application of the numeration theorem 3.1 we give a way of defining order types as sets. We cannot prove the class $\{x \mid x \approx b\}$ (cf. V, § 2) to be represented by a set, but can assign to every order b a set $\mathrm{o}(b)$ in such a way that for every order d

$$\mathrm{o}(b) = \mathrm{o}(d) \leftrightarrow b \approx d.$$

Indeed by V, Df 2.6, V, Df 2.7 and V, 2.10 we have

$$\mathrm{Or}(b) \,\&\, \Delta_1 b^* \underset{c*}{\sqsupset} a^* \rightarrow \mathrm{Or}(\breve{c} \mid b \mid c, a) \,\&\, \breve{c} \mid b \mid c \approx b$$

and by 3.5

$$(Ex)(\Delta_1 b^* \underset{x*}{\sqsupset} (\aleph(\Delta_1 b))^*,$$

which together give

$$[1] \qquad \mathrm{Or}(b) \rightarrow (Ez)(z \approx b \,\&\, \Delta_1 z = \aleph(\Delta_1 b)).$$

On the other hand by V, 2.13 and 3.6 we have

$$[2] \qquad b \approx d \rightarrow \aleph(\Delta_1 b) = \aleph(\Delta_1 d).$$

From [1], [2] and V, 2.14 we get

3.13 $$\mathrm{Or}(b) \,\&\, \mathrm{Or}(d) \cdot\!\rightarrow\!\cdot\, b \approx d \leftrightarrow \{x \mid x \approx b \,\&\, \Delta_1 x = \aleph(\Delta_1 b)\} \equiv$$
$$\equiv \{x \mid x \approx d \,\&\, \Delta_1 x = \aleph(\Delta_1 d)\}.$$

By 3.13 is expressed that the order type of an order b can be characterized by the class $\{x \mid x \approx b \,\&\, \Delta_1 x = \aleph(\Delta_1 b)\}$. This class however is represented by a set, since it is a subclass of the class of all orders of $\aleph(\Delta_1 b)$ which again is a subclass of the set $\pi(\aleph(\Delta_1 b) \times \aleph(\Delta_1 b))$. Hence defining

Df 3.4 $$\mathrm{o}(b) = \iota_z(z^* \equiv \{x \mid x \approx b \,\&\, \Delta_1 x = \aleph(\Delta_1 b)\}$$

we have

3.14 $$\mathrm{Or}(b) \,\&\, \mathrm{Or}(d) \cdot\!\rightarrow\!\cdot\, \mathrm{o}(b) = \mathrm{o}(d) \leftrightarrow b \approx d.$$

VI, § 4. Zorn's lemma and related principles

In newer algebra there is a tendency of avoiding direct applications of the axiom of choice and also of the wellorder theorem. Instead some general maximum principles are used which make no explicit reference to wellorder. We here consider some of these principles and show their equivalence. We start from the following principle stated by F. Hausdorff [1914].

4.1 $$\mathrm{Po}(c) \rightarrow (Ex)(\mathrm{Ch}(x, c) \,\&\, (y)(x \subseteq y \,\&\, \mathrm{Ch}(y, c) \rightarrow x = y)).$$

«Every partial order has a subset which is a maximal chain».

In order to prove 4.1 we proceed by the method of Zermelo's first proof of the wellorder theorem, applied already to the second proof of the theorem of generated numeration (IV, § 3).

Firstly we apply the axiom of choice in the following way: For every subset b of $\Delta_1 c$ there exists the class

$$\{x \mid x \in (\Delta_1 c)^{-}b \;\&\; \mathrm{Or}(c \cap (b; x \times b; x))\}$$

– briefly $\{x \mid \mathfrak{C}(x, b)\}$ – which of course, being a subclass of $\Delta_1 c$ is represented by a set. Note that this set is non void only when $b \subset \Delta_1 c$ and $\mathrm{Or}(c \cap (b \times b))$.

Further then there exists the function

$$[1] \qquad \{uv \mid u \subseteq \Delta_1 c \;\&\; v^* \equiv \{x \mid \mathfrak{C}(x, u)\} \;\&\; v \neq 0\}.$$

This function again is represented by a functional set f since its domain is a subclass of $\pi(\Delta_1 c)$; f fulfils the condition of 2.3 with m being $\Delta_1 f$, hence we have the existence of a function g such that

$$\Delta_1 g = \Delta_1 f \;\&\; (x)(x \in \Delta_1 f \rightarrow g`x \in f`x).$$

g assigns to every subset b of $\Delta_1 c$ which is ordered by c, provided there is an element e of $\Delta_1 c^{-}b$ such that $b; e$ is again ordered by c, one of these elements e.

Now we consider the class N of the numerations of subsets of $\Delta_1 c$, generated by g^*. By IV, 3.2 we have that $\bigcup N$ is represented by a set s which is a numeration of a subset t of $\Delta_1 c$, again generated by g^*.

But t determines in c a chain; in order to show this it is sufficient[1]) to prove that for any two different elements d, e of t we have $\langle d, e\rangle \in c \vee \langle e, d\rangle \in c$. Let e be that one of them which by the numeration of t is assigned to the higher ordinal; then since the numeration of t is generated by g^*, e is the value assigned by g to the set q of those elements of t preceding e in the numeration, and therefore by the characterizing property of g the set $(c \cap (q; e \times q; e))$ is a chain, but d and e are elements of $q; e$ and so indeed $\langle d, e\rangle \in c \vee \langle e, d\rangle \in c$. Thus we have that t determines a chain $c \cap (t \times t)$ of c.

[1]) We are relying here of course on our premise Po(c).

Now this chain is also a maximal chain of c, for if there were a chain of which $c \cap (t \times t)$ is a proper subset and r an element of $\Delta_1 c$ not in t, then $c \cap (t;r \times t;r)$ would be a chain. So we should have $r \in f`t$, therefore $t \in \Delta_1 f$ and also $t \in \Delta_1 g$. If now $p = g`t$, then $p \notin t$; on the other hand, if m is the domain of the numeration s, then $s; \langle m, p \rangle$ would be again a numeration of a subset of $\Delta_1 c$ generated by g^*, and so, by the definition of s, p would have to belong to t.

From this general theorem 4.1, the Hausdorff principle, we draw the Zorn lemma [1935] by considering the partial order of the elements of a set by the subset relation. In fact for every set a the set representing the class

$$\{xy \mid x \in a \;\&\; y \in a \;\&\; x \subseteq y\}$$

is a partial order of a. By the Hausdorff principle there is in this partial order a maximal chain c. If $\sum \Delta_1 c$ is an element of a it must be a *maximal element*, that is an element which is not a proper subset of an other element. For, if there were an element b of a such that $\sum \Delta_1 c \subset b$ then every element of $\Delta_1 c$ were a proper subset of b and so the chain c could be extended.

The condition here occurring that $\sum \Delta_1 c \in a$ is certainly then satisfied if for every subset s of a which with respect to the subset-relation is the domain of a chain [1]), $\sum s \in a$. A set satisfying this condition will be called closed, formally

Df 4.1 $$\mathrm{Cl}(a) \leftrightarrow (x)(x \subseteq a \;\&\; (u)(v)(u \in x \;\&\; v \in x \rightarrow \\ \rightarrow u \subseteq v \vee v \subseteq u) \rightarrow \sum x \in a).$$

So we come to the statement of the Zorn lemma: Every closed set has a maximal element; formally

4.2 $$\mathrm{Cl}(a) \rightarrow (Ex)(x \in a \;\&\; (z)(z \in a \rightarrow \overline{x \subset z})).$$

In many applications an other similar principle, first stated by Teichmüller [1939] is especially handy. For its easier formulation we call a non empty class A *characterized by its finite elements*

---

[1]) In usual mathematics we should say: "... which with respect to the subsetrelation is a chain." The more complicated formulation is here necessary, because we understand by a chain an order and not an ordered set.

if a set c belongs to A if and only if every finite subset of c belongs to A.

Df 4.2 $$\mathrm{Fch}(A) \leftrightarrow A \neq \Lambda \ \& \ (x)(x \in A \leftrightarrow \leftrightarrow (y)(y \subseteq x \ \& \ \mathrm{Fin}(y) \rightarrow y \in A)).$$

The definition of Fch(a) is quite corresponding.

The Teichmüller principle now says that if a class of subsets of a set is characterized by its finite elements, it has a maximal element.

4.3 $$\mathrm{Fch}(A) \ \& \ (x)(x \in A \rightarrow x \subseteq m) \cdot\!\rightarrow\!\cdot (Ey)(y \in A \ \& \ \& \ (z)(z \in A \rightarrow \overline{y \subset z})).$$

In virtue of A 4 we can replace 4.3 by the even more simple statement

4.4 $$\mathrm{Fch}(a) \rightarrow (Ey)(y \in a \ \& \ (z)(z \in a \rightarrow \overline{y \subset z}))$$

«Every set which is characterized by its finite elements has a maximal element.»

Indeed 4.3 follows from 4.4 since

$$(x)(x \in A \rightarrow x \subseteq m) \rightarrow A \subseteq \pi(m)^*$$

and

$$\mathrm{Fch}(A) \ \& \ a^* \equiv A \rightarrow \mathrm{Fch}(a).$$

On the other hand 4.3 entails 4.4 since for every set a we have

$$(x)(x \in a \rightarrow x \subseteq \textstyle\sum a)$$

i.e. every element of a is a subset of the sum of the elements of a.

In order to derive 4.4 from the Zorn lemma 4.2 we have to prove

4.5 $$\mathrm{Fch}(a) \rightarrow \mathrm{Cl}(a)$$

i.e. $$a \neq 0 \ \& \ (x)(x \in a \leftrightarrow (y)(y \subseteq x \ \& \ \mathrm{Fin}(y) \rightarrow y \in a)) \rightarrow (x)(x \subseteq a \ \& \ \& \ (u)(v)(u \in x \ \& \ v \in x \rightarrow u \subseteq v \vee v \subseteq u) \rightarrow \textstyle\sum x \in a).$$

The idea of the proof is the following. Let b be a subset of a such that $(u)(v)(u \in b \ \& \ v \in b \rightarrow u \subseteq v \vee v \subseteq u)$. We have to show that $\sum b \in a$. By Fch(a) it is sufficient for this to show that every finite subset s of $\sum b$ is an element of a[1]. Every element e_i of s

[1] We can omit here the case $s = 0$, since, as easily seen, $\mathrm{Fch}(a) \rightarrow 0 \in a$.

is an element of some element of b; thus for each of the finitely many elements e_i of s there is an element c_i of b such that $e_i \in c_i$. From the noted property of b follows that there is one of the c_i of which all the others are subsets. If c is this element of b, then every element of s is an element of c, hence $s \subseteq c$. Besides since $b \subseteq a$, we have $c \in a$ and so in virtue of $\text{Fch}(a)$, because s is finite, we have in fact $s \in a$. For formally carrying out this proof we have to procede by a numeral induction with respect to the multitude of s.

From Teichmüllers principle 4.3 we come back to the axiom of choice A 5 in the following way: Let a be a set of pairs and A be the class $\{x \mid x \subseteq a \;\&\; \text{Ft}(x)\}$, then obviously $\text{Fch}(A)$ – since a set of pairs is a function if and only if every subset consisting of two elements is a function. Therefore by 4.3, A has a maximal element f. It only remains to show that $\Delta_1 f = \Delta_1 a$. But in the other case there would exist a pair $\langle d, e \rangle \in a$ such that $d \notin \Delta_1 f$, and $f; \langle d, e \rangle$ would be a function and also a subset of a contrary to the maximum property of f.

Thus we have found that the principles of Hausdorff 4.1, Zorn 4.2 and Teichmüller 4.3, 4.4 are all, on the supposition of A 4, equivalent with the axiom of choice.

Attention still may be called to the possibility of obtaining directly the numeration theorem 3.1 by application of Zorn's lemma: by 3.11 the class of numerations of subsets of a set a is represented by a set c. Now this set c is closed. Indeed of any two numerations of subsets of a, s and t, such that $s \subseteq t \vee t \subseteq s$ one is a segment of the other and from this together with III, 2.4 it follows that the sum of every chain (with respect to the subset-relation) of elements of c is again an element of c. Therefore by 4.2, c has a maximal element p. Now the subset of a of which p is a numeration must be a itself. For if there were an element d of a which is not in the converse domain of p then we should have $p; \langle \Delta_1 p, d \rangle \in c$ and so p would not be maximal.

Here it particularly appears that the rôle of the Zorn lemma consists not only in replacing applications of A5 but also in sparing the reasonings to be used for proving the theorem of generated numeration, either by Zermelo's method or with the general recursion theorem; the same holds of the Hausdorff and Teichmüller principle. We mention here an application of this kind of 4.4 which yields a general theorem, stated by R. Büchi [1953].

For its formulation we denote by $\mathfrak{C}(c, r, a)$ the expression

$$c \subseteq a \;\&\; (x)(y)(x \neq y \;\&\; x \in c \;\&\; y \in c \rightarrow \langle x, y\rangle \in r \vee \langle y, x\rangle \in r).$$

«c is an r-connected subset of a.»

Then the formal statement is given by

4.6 $$r \subseteq a \times a \;\&\; \mathrm{Ft}(f) \;\&\; (x)(\mathfrak{C}(x, r, a) \rightarrow x \in \Delta_1 f \;\&\; (z)(z \in x \rightarrow \rightarrow \langle z, f`x\rangle \in r)) \cdot\!\rightarrow\!\cdot (Eu)(\mathfrak{C}(u, r, a) \;\&\; f`u \in u).$$

This formula on the one hand can be proved analogously as the theorem of generated numeration IV, 3.1 by considering upon the premise the class T of numerations s of subsets of a, such that, if $n \in \Delta_1 s$ and $b = \Delta_2 sg(s, n)$, then $\mathfrak{C}(b, r, a) \;\&\; s`n = f`b$. This class, by 3.11, is represented by a set t and one proves $\mathfrak{C}(\Delta_2 \sum t), r, a) \;\&$ $\& \; f`\Delta_2 \sum t \in \Delta_2 \sum t$.

Here in virtue of the premise on f, no application of the axiom of choice is needed.

But the proof becomes much simpler by using Teichmüller's principle. Indeed for given a and r, $r \subseteq a \times a$; let H be the class of sets c such that $\mathfrak{C}(c, r, a)$, which by A 4 is represented by a set h. Then as easily seen, we have $\mathrm{Fch}(h)$ and therefore by 4.4 there exists a maximal element p in h. Hence in virtue of the premise of 4.6 we must have $f`p \in p$, since $\mathfrak{C}((p; f`p), r, a)$.

VI, § 5. Axiom of infinity. Denumerability

The general theorems for which we used in the last sections the axioms A 4 and A 5 are intended properly for non finite sets. If we have to deal only with finite sets, then the application of A 4 and A 5 can be dispensed with. In fact by the theorems of finite sets III, § 5 we can prove that the class of subsets of a finite set is represented by a set, and by numeral induction it is provable that for every set of pairs a with a finite domain there exists a subset with the same domain which is a function.

On the other hand till now the existence of a non finite set cannot be demonstrated. This in the foregoing was not sensible, since for number theory and the general theory of ordinals we need not the existence of non finite sets. However for analysis and for the Cantor set theory the dealing with non finite sets is essential. In order to enable it, we have to introduce an axiom of infinity.

There are various possible forms of stating this axiom. The form we choose is related to Cantor's way of starting in his set theory from the set of natural numbers. This set in our system must be an ordinal which has as elements just the natural numbers and hence is the least ordinal higher than every natural number. We denote it as usual by ω and thus state the axiom of infinity by the formula

A 6 $$a \in \omega \leftrightarrow \mathrm{Nu}(a).$$

This formula is obviously equivalent with

5.1 $$\omega^* \equiv \{x \mid \mathrm{Nu}(x)\}.$$

As immediate consequences we note that every class of numerals is represented by a set:

5.2 $$A \subseteq \{x \mid \mathrm{Nu}(x)\} \rightarrow \mathrm{Rp}(A).$$

Let us regard in what respect A 6 has the signification of an axiom of infinity. We have not yet defined infinity, but only still finiteness. Hence the simplest and most natural definition of infinity here is

Df 5.1 $$\begin{cases} \mathrm{Infin}(a) \leftrightarrow \overline{\mathrm{Fin}(a)} \\ \mathrm{Infin}(A) \leftrightarrow \overline{\mathrm{Fin}(A)}. \end{cases}$$

This purely negative characterization of infinity is equivalent with the following positive one

5.3 $$\mathrm{Infin}(A) \leftrightarrow (x)(\mathrm{Nu}(x) \rightarrow (Ey)(y^* \subseteq A \;\&\; y \sim x).$$

Namely on one hand as a consequence of III, 5.6 we obtain

[1] $$\mathrm{Nu}(n) \;\&\; n \sim a \;\&\; b \subseteq a \rightarrow \overline{b \sim n'},$$

and so it follows the implication from the right to the left of 5.3. On the other hand by numeral induction with respect to n we get

[2] $$\mathrm{Infin}(A) \;\&\; \mathrm{Nu}(n) \rightarrow (Ex)(x \sim n \;\&\; x^* \subseteq A).$$

From 5.3 together with A 6 we obtain Infin(ω) and thus

5.4 $$(Ex)\ \mathrm{Infin}(x).$$

We also can come back from 5.4 to A 6. In fact we have

5.5 $$\mathrm{Infin}(a) \rightarrow \mathrm{Rp}(\{x \mid \mathrm{Nu}(x)\}).$$

For the proof we consider the class of pairs

$$\{xy \mid x \subseteq a \,\&\, \mathrm{Fin}(x) \,\&\, y = \mathrm{mlt}(x)\}.$$

This class is a function and by A 4 its domain is represented by a set, and hence also its converse domain is represented. But this converse domain, in virtue of our premise and [2] is $\{x \mid \mathrm{Nu}(x)\}$; thus $\{x \mid \mathrm{Nu}(x)\}$ is represented [1].

An other, very simple proof of 5.5 is by means of the numeration theorem but without A 4. Indeed the domain of a numeration of an infinite set a must be an infinite ordinal and thus have $\{x \mid \mathrm{Nu}(x)\}$ as subclass.

We further easily come from A 6 to other statements of the axiom of infinity. Zermelo's axiom of infinity says

5.6 $$(Ex)(0 \in x \,\&\, (y)(y \in x \rightarrow [y] \in x)).$$

This is obtained from A 6 by considering the iterator of the function $\{xy \mid y = [x]\}$ on 0. By III, 4.1 $\mathsf{J}(\{xy \mid y = [x]\}, 0)$ is a function with the domain $\{x \mid \mathrm{Nu}(x)\}$. By A 6 and II, 3.8 the converse domain is represented by a set c, and from the property of the iterator follows

$$0 \in c \,\&\, (y)(y \in c \rightarrow [y] \in c),$$

so that we get 5.6.

The inverse passage from 5.6 to A 6 is also possible. Namely first by numeral induction we prove

$$0 \in c \,\&\, (y)(y \in c \rightarrow [y] \in c) \quad \rightarrow \quad \Delta_2\mathsf{J}(\{xy \mid y = [x]\}, 0) \subseteq c^*.$$

This together with 5.6 and II, 2.1 gives

[1] $$\mathrm{Rp}(\Delta_2\mathsf{J}(\{xy \mid y = [x]\}, 0)).$$

On the other hand by III, 4.2 we have

[2] $$\mathrm{Crs}(\mathsf{J}(\{xy \mid y = [x]\}, 0);$$

[1] and [2] yield by II, 3.8

[3] $$\mathrm{Rp}(\Delta_1(\mathsf{J}(\{xy \mid y = [x]\}, 0)),$$

[1]) The proof in this form was given by K. Gödel [1940].

and by III, 4.1

$$\Delta_1 \mathsf{J}(\{xy \mid y = [x]\}, 0) \equiv \{x \mid \mathrm{Nu}(x)\},$$

so that

$$\mathrm{Rp}(\{x \mid \mathrm{Nu}(x)\}).$$

An other form of the axiom of infinity arises from Dedekinds definition of infinity. According to this definition the assertion of the existence of an infinite set says that there exists a set which has a one-to-one correspondence to a proper subset

5.7 $$(Ex)(Ey)(y \subset x \ \&\ x \sim y).$$

This in an obvious way follows from A 6 since, even in many ways, we have a one-to-one correspondence of ω to a proper subset of it.

For the passage from 5.7 to A 6 let a be a set having a one-to-one correspondence c with a proper subset b; then we have $(x)(x \in a \rightarrow c`x \in b)$. If now $d \in a^{-}b$, then the iterator $\mathsf{J}(c^*, d)$, by III, 4.2 is a one-to-one correspondence; its domain is $\{x \mid \mathrm{Nu}(x)\}$ and its converse domain is a subclass of a, and thus represented by a set, hence also $\{x \mid \mathrm{Nu}(x)\}$ is represented.

A pregnant form of an axiom of infinity is also that which occurs in von Neumann's system of set theory and which was adopted by Gödel

5.8 $$(Ex)(x \neq 0 \ \&\ (y)(y \in x \rightarrow (Ez)(z \in x \ \&\ y \subset z))).$$

«There is a nonempty set which has no maximal element».

The proof of 5.8 from A 6 is immediate since ω is a non empty set and has no maximal element. For the inverse passage we may use either A 5 or A 4. The reasoning by A 5 is as follows: Let a be a non empty set having no maximal element. Applying A 5 to the class (represented by a set)

$$\{xy \mid x \in a \ \&\ y \in a \ \&\ x \subset y\}$$

we infer the existence of a function f assigning to every element c of a an other element of a of which c is a proper subset. The iterator of this function on some element of a is a one-to-one correspondence between $\{x \mid \mathrm{Nu}(x)\}$ and a subclass of a. Namely the element assigned to the lower of two numerals is a proper subset of the other. Now the converse domain of this iterator, as a subclass of a is represented by a set, hence also the domain $\{x \,|\, \mathrm{Nu}(x)\}$ is represented.

The proof by A 4 goes as follows [1]: From the assumption that the non-empty set a has no maximal element it can be inferred by numeral induction that for every natural number m there exists a functional set f such that

$$\Delta_1 f = m \;\&\; \Delta_2 f \subseteq a \;\&\; (x)(x' \varepsilon m \rightarrow f(x) \subset f(x')).$$

As this function can be shown to be a one-to-one correspondence, it follows by Df 5.3 (III, § 5): $\mathrm{mlt}(\Delta_2 f) = m$. Hence, if $\mathfrak{G}$ is the function assigning to every finite subset of a its multitude, then $\Delta_2\mathfrak{G} \equiv \{x \,|\, \mathrm{Nu}(x)\}$. Now, by A 4, $\Delta_1\mathfrak{G}$ is represented by a set, and therefore, by the theorem of replacement, $\Delta_2\mathfrak{G}$, i.e. $\{x \,|\, \mathrm{Nu}(x)\}$ is represented.

In a certain relatedness to the proofs of equivalences between axioms of infinity are those proofs which are about equivalences between definitions of infinity or, what comes out to the same, between definitions of finiteness. As already mentioned (III, § 5) proofs of this kind have been studied in particular with respect to the need of applying the axiom of choice. Here we take the opportunity of mentioning two elegant definitions of finiteness, connected with the concept of order and wellorder.

5.9 $$\mathrm{Fin}(a) \leftrightarrow (x)(\mathrm{Or}(x, a) \rightarrow \mathrm{Wor}(x, a))$$

«A set is finite if and only if every order of it is a wellorder».

5.10 $$\mathrm{Fin}(a) \leftrightarrow (x)(y)(\mathrm{Or}(x, a) \;\&\; \mathrm{Or}(y, a) \rightarrow x \approx y).$$

The implications from left to right in 5.9 and 5.10 are both derivable with the help of III, 5.3, V, Df 2.6, V, Df 2.7, V 3.5 and the formula

5.11 $$\mathrm{Fin}(a) \;\&\; \mathrm{Or}(b, a) \rightarrow (Ex)(\mathrm{Num}(x, a) \;\& \\ \& \; \Delta_1 x = \mathrm{mlt}(a) \;\&\; b = \breve{x} \mid \mathrm{no}(\Delta_1 x) \mid x)$$

which is obtainable by means of a numeral induction with respect to $\mathrm{mlt}(a)$. The proof of the inverse implications can be made with the help of the wellorder theorem 3.3, using the theorem that

[1] This proof, conformous to that one in [Bernays 1942] p. 68, is a slight modification of the proof given by Gödel in [Gödel 1940] 8.51, pp. 32–33.

a wellorder, whose inverse order is also a wellorder, is an order of a finite set, formally

5.12 $$\mathrm{Wor}(b, a) \;\&\; \mathrm{Wor}(\breve{b}) \rightarrow \mathrm{Fin}(a).$$

This statement in virtue of the connection between wellorder and numeration (V, 3.6) comes back to the assertion

5.13 $$\mathrm{Od}(n) \;\&\; \mathrm{Wor}(\overset{\smile}{\mathrm{no}(n)}) \rightarrow \mathrm{Nu}(n)$$

which almost directly follows from the definitions of the concepts Wor, no, and Nu and the general theorems on ordinals.

From A 6 immediately follows that ω is the lowest infinite ordinal. This in particular entails that ω is an initial number of a Zahlenklasse, and thus its own cardinal number. The sets whose cardinal number is ω are called *denumerable*. This comes out to define

Df 5.2 $$\mathrm{Denum}(a) \leftrightarrow a \sim \omega.$$

We further call a class denumerable if it is represented by a denumerable set

Df 5.3 $$\mathrm{Denum}(A) \leftrightarrow (Ex)(x^* \equiv A \;\&\; x \sim \omega).$$

According to this we have by the theorem of replacement

5.14 $$A \ulcorner\!C\!\urcorner \{x \mid \mathrm{Nu}(x)\} \rightarrow \mathrm{Denum}(A) \;\&\; \mathrm{Rp}(A).$$

Concerning denumerability there are some often used theorems. The first says that every infinite set has a denumerable subset.

5.15 $$\mathrm{Infin}(a) \rightarrow (Ex)(x \subseteq a \;\&\; \mathrm{Denum}(x)).$$

The most simple proof is by using that there exists a numeration s of a. Since a is infinite, the domain of s cannot be a natural number. So it must be an infinite ordinal and ω is a subset of it. Let b be the subset of the converse domain of s, consisting of those elements of a, which in s are assigned to a natural number; then b is a denumerable subset of a.

5.15 in particular can be used for an alternative proof of the implications from the right to the left of the formulas 5.9 and 5.10. This goes, using contraposition, in virtue of the circumstance that the inverse order of the natural order of ω is not a wellorder.

As a further assertion on denumerability we have

5.16 $\text{Denum}(a)\ \&\ b \subseteq a \rightarrow \text{Fin}(b) \vee \text{Denum}(b)$.

«Every subset of a denumerable set is finite or denumerable».

Namely a subset b of a denumerable set is of equal power with a subset of ω. Therefore we have a numeration of b by the natural order of that subset of ω. Let m be the domain of this numeration, then by V, 3.10 either m is lower than ω and then it is a natural number and b is finite, or $m = \omega$ and b is denumerable.

A direct application of 5.16 concerns *ω-sequences*. By an ω-sequence we understand a sequence with the domain ω.

Df 5.4 $\text{Dsq}(s) \leftrightarrow \text{Sq}(s)\ \&\ \Delta_1 s = \omega$.

The assertion in question is that the converse domain of an ω-sequence is finite or denumerable.

5.17 $\text{Dsq}(s) \rightarrow \text{Fin}(\Delta_2 s) \vee \text{Denum}(\Delta_2 s)$.

Note that an ω-sequence can contain repetitions of members. By 2.8 there exists a one-to-one correspondence between $\Delta_2 s$ and a subset of ω, which by 5.16 is finite or denumerable. Of course, the one-to-one correspondence could here also be obtained with the help of the principle of least ordinal number, without applying the axiom of choice which is used for 2.8.

As a corollary of 5.17 we have

5.18 $\text{Ft}(f)\ \&\ \text{Denum}(\Delta_1 f) \rightarrow \text{Fin}(\Delta_2 f) \vee \text{Denum}(\Delta_2 f)$.

Namely composing the one-to-one correspondence between ω and $\Delta_1 f$ with f we get a denumerable sequence with the converse domain $\Delta_2 f$, and 5.17 can be applied.

Next we state

5.19 $\text{Denum}(a)\ \&\ \text{Denum}(b) \rightarrow \text{Denum}(a \times b)$.

In order to prove this it is sufficient to give a one-to-one correspondence between $\omega \times \omega$ and ω. There are known many such one-to-one correspondences. For instance a one-to-one correspondence between the pairs $\langle m, n\rangle$ of natural numbers and the natural number k is defined by $2^m(2n+1) = k+1$, in the usual arithmetic notation.

Still we have the theorem

5.20 $\text{Dsq}(s)\ \&\ (z)(z \in \Delta_2 s \rightarrow \text{Denum}(z)) \rightarrow \text{Denum}(\sum_x(\omega, s^{\prime}x))$

«The sum of the members of an ω-sequence of denumerable sets is again denumerable».

For a more handy formulation of the proof we introduce the concept of a denumeration of a set a, defining it as a one-to-one correspondence between ω and a. Now first we are to show that there exists a functional set g assigning to every natural number k a denumeration of $s`k$. We consider the class of pairs $\langle u, v\rangle$ where $u$ is a natural number and $v$ is the set representing the class of denumerations of $s`u$, which indeed is represented, as a consequence of 3.11. The said class of pairs is a function whose domain is represented by ω and is represented by a functional set f. Applying to f the axiom of choice [1]) in the form 2.3 with m being ω, we infer the existence of a functional set g such that

$$\Delta_1 g = \omega \ \& \ (x)(x \in \omega \rightarrow \mathrm{Num}(g`x, s`x) \ \& \ \Delta_1 g`x = \omega).$$

Thus g is a function such as has been stated to exist. Now let H be the function

$$\{xy \,\|\, (Eu)(Ev)(x = \langle u, v\rangle \ \& \ \langle u, v\rangle \in \omega \times \omega \ \& \ y = (g`u)`v)\},$$

then we have

$$\Delta_1 H \equiv (\omega \times \omega)^*$$

and

$$\Delta_2 H \equiv \sum_x (\omega, s`x)^*.$$

From the first also follows that H is represented by a functional set h. Since $\omega \times \omega$ is denumerable by 5.19, we have by 5.18 that $\Delta_2 h$ is denumerable or finite. But the case that it is finite is excluded since every member of s is supposed to be denumerable.

An obvious application of the axiom A 6 yields that for every function G with $\Delta_1 G \equiv \{x \,\|\, \mathrm{Nu}(x)\}$ the converse domain $\Delta_2 G$ and also $\bigcup \Delta_2 G$ is represented, as follows by II, 3.8 and II, 3.7. This holds in particular for any numeral iterator, which indeed by III, 4.1 is such a function. By this way it also results in virtue of III, 4.10 that the transitive closure of any set is represented, so that we have

5.21 $$\mathrm{Rp}(\overline{|a^*|}).$$

---

[1]) Concerning alternatives to 5.20 arising in absence of the axiom of choice cf. A. Church [1927] and E. Specker [1957].

CHAPTER VII

ANALYSIS; CARDINAL ARITHMETIC; ABSTRACT THEORIES

VII, § 1. Theory of real numbers

Going on to discuss the application of our full system of axioms A 1–A 6 to classical mathematics, we shall consider in particular two domains: classical analysis and Cantor's theory of powers.

In order to show the possibility of embodying analysis in our set theoretic frame, it will be sufficient to indicate the way of procedure for constructing the system of real numbers and proving the laws of computation and the continuity property, as also for handling with functions of real variables. We shall abstain here from a more detailed treatment which gives no difficulty on principle.

From our discussion it will result, as it was likewise shown in [Bernays 1942] that for obtaining classical analysis, also with its impredicative procedures, the application of axiom A 4 can be avoided, since the axiom A 6 already implies that every denumerable class is represented. Even for a more liberal delimitation of the frame the full axiom A 4 is not needed but instead, as we shall see, a strengthening of axiom A 6 is sufficient, namely an axiom postulating that the class of all number sets, (in place of the class of natural numbers), is represented.

In general the construction of the system of real numbers, starting from the series of natural numbers, is made in several steps, by first constructing the system of rationals which themselves are defined as sets of signed fractions. The stage of the rational numbers can here be dispensed of; and besides we have a direct passage from natural numbers to signed fractions by characterizing these as fraction triplets.

By a fraction triplet we understand a set $\langle\langle a, b\rangle, c\rangle$ where a, b are arbitrary natural numbers, and c is a natural number different from 0.

Df 1.1 $\quad \mathrm{Ftp}(d) \leftrightarrow (Ex)(Ey)(Ez)(\mathrm{Nu}(x) \;\&\; \mathrm{Nu}(y) \;\&\; \mathrm{Nu}(z) \;\&\; z \neq 0 \;\&\; \\ \&\; d = \langle\langle x, y\rangle\, z\rangle).$

We use the word "fraction triplet" in view of the arithmetic rôle to be given to these triplets according to the interpretation that $\langle\langle a, b\rangle, c\rangle$ means $\frac{a-b}{c}$ (as usually denoted).

In conformity with this interpretation we define the sum

Df 1.2 $\quad p +^{r} q = \iota_t(\mathrm{Ftp}(p) \;\&\; \mathrm{Ftp}(q) \;\&\; \\ \&(x)(y)(z)(u)(v)(w)(p = \langle\langle x, y\rangle, z\rangle \;\&\; q = \langle\langle u, v\rangle, w\rangle \rightarrow \\ \rightarrow t = \langle\langle (w\cdot x) + (z\cdot u), (w\cdot y) + (z\cdot v)\rangle, z\cdot w\rangle)),$

so that

1.1 $\quad \mathrm{Ftp}(\langle\langle a, b\rangle, c\rangle) \;\&\; \mathrm{Ftp}(\langle\langle k, l\rangle, m\rangle) \cdot\!\rightarrow\!\cdot \\ \cdot\!\rightarrow\!\cdot \langle\langle a, b\rangle, c\rangle +^{r} \langle\langle k, l\rangle, m\rangle = \langle\langle (m\cdot a) + (c\cdot k), (m\cdot b) + (c\cdot l)\rangle, c\cdot m\rangle$

becomes provable. Correspondingly the difference and the product, symbolized by $-^{r}$ and $\cdot^{r}$, are defined so that we have

1.2 $\quad \mathrm{Ftp}(\langle\langle a, b\rangle, c\rangle) \;\&\; \mathrm{Ftp}(\langle\langle k, l\rangle, m\rangle) \cdot\!\rightarrow\!\cdot \\ \cdot\!\rightarrow\!\cdot \langle\langle a,b\rangle,c\rangle -^{r} \langle\langle k,l\rangle,m\rangle = \langle\langle (m\cdot a) + (c\cdot l), (m\cdot b) + (c\cdot k)\rangle, c\cdot m\rangle \;\&\; \\ \&\; \langle\langle a, b\rangle, c\rangle \cdot^{r} \langle\langle k, l\rangle, m\rangle = \langle\langle (a\cdot k) + (b\cdot l), (a\cdot l) + (b\cdot k)\rangle, c\cdot m\rangle.$

Further in accordance with our interpretation of fraction triplets we define the predicates concerning their sign, equality and order. A fraction triplet $\langle\langle a, b\rangle, c\rangle$ is called *positive*, *negative* or a *nulltriplet* according as $b \in a$ or $a \in b$ or $a = b$; and a fraction triplet p is called *greater than*, *less than*, or *equally great* as the fraction triplet q according as $p -^{r} q$ is positive, negative or a nulltriplet. The formal definition of these concepts can be given as follows

Df 1.3 $\quad p =^{r} q \leftrightarrow \mathrm{Ftp}(p) \;\&\; \mathrm{Ftp}(q) \;\&\; (Ex)(Ey)(Ez)(Eu)(Ev)(Ew) \\ (p = \langle\langle x, y\rangle, z\rangle \;\&\; q = \langle\langle u, v\rangle, w\rangle \;\&\; \\ \&\; (w\cdot x) + (z\cdot v) = (w\cdot y) + (z\cdot u)).$

Df 1.4 $\quad p <^{r} q \leftrightarrow \mathrm{Ftp}(p) \;\&\; \mathrm{Ftp}(q) \;\&\; (Ex)(Ey)(Ez)(Eu)(Ev)(Ew) \\ (p = \langle\langle x, y\rangle, z\rangle \;\&\; q = \langle\langle u, v\rangle, w\rangle \;\&\; \\ \&\; ((w\cdot x) + (z\cdot v)) \in ((w\cdot y) + (z\cdot u))).$

Df 1.5 $\quad p >^{r} q \leftrightarrow q <^{r} p.$

The properties of p being a positive, or a negative fraction triplet or a nulltriplet can now be expressed by

1.3 $$p >^r \langle\langle 0, 0\rangle, 1\rangle,\ p <^r \langle\langle 0, 0\rangle, 1\rangle,\ p =^r \langle\langle 0, 0\rangle, 1\rangle.$$

A fraction triplet p is called a 1-triplet if $p =^r \langle\langle 1, 0\rangle, 1\rangle$. Note that a fraction triplet $\langle\langle a, b\rangle, c\rangle$ is a 1-triplet if and only if $a = b + c$.

To every positive fraction triplet there is an equally great one of the form $\langle\langle a, 0\rangle, c\rangle$, and to every negative an equally great one of the form $\langle\langle 0, b\rangle, c\rangle$. In this way we get the connection between fraction triplets and fractions; indeed we can define for $c \neq 0$, $\frac{a}{c}$ to be $\langle\langle a, 0\rangle, c\rangle$ and $-\frac{b}{c}$ to be $\langle\langle 0, b\rangle, c\rangle$.

Our definition of "equally great" ensures that for all the introduced operations and relations the substitutivity of equally great fraction triplets holds. Further now it can be directly stated that all the laws of an ordered field are satisfied for the fraction triplets, in particular the existence of a multiplicative inverse for every fraction triplet which is not a nulltriplet.

Let us notice also that the class of fraction triplets is represented by a set since it is a subclass of the set $(\omega \times \omega) \times \omega$; by the same reason every class of fraction triplets is represented by a set. Every such set is also either finite or denumerable, as follows from VI, 5.16 and 5.19.

Now we come to define real numbers as special sets of fraction triplets by the method of Dedekind, understanding by a real number a non void proper initial section without greatest element in the ordered set of fraction triplets.

Df 1.6 $$\mathrm{Re}(c) \leftrightarrow c^* \subset \{x \,|\, \mathrm{Ftp}(x)\} \,\&\, c \neq 0 \,\&\, (x)(x \in c \rightarrow (Ey)(y \in c \,\&\, \\ \&\, x <^r y)) \,\&\, (x)(y)(y \in c \,\&\, (x =^r y \vee x <^r y) \rightarrow x \in c).$$

According to the given definition of real number, as well known, equality of real numbers is simply set theoretic equality, and the ordering relation $p \leqq q$ is directly the subset relation. A real number is positive if it has some positive fraction triplet as an element; it is negative if there is some negative fraction triplet which is not an element of it. The real number null is the set of all negative fraction triplets. Obviously thus every real number is either positive or negative or null.

A real number p is called *rational* if there exists a fraction triplet t such that p represents the class $\{x \mid x <^r t\}$. In the case that p is positive or null, the fraction triplet t in question can be chosen in the form $\langle\langle k, 0\rangle, l\rangle$, so that p is the set representing the class $\{x \mid x <^r \langle\langle k, 0\rangle, l\rangle\}$. We denote this set by $\left[\frac{k}{l}\right]$. At once we define $[k] = \left[\frac{k}{1}\right]$. According to this in particular $[0]$ is the real number null.

Now we are defining the arithmetic operations on real numbers. By the arithmetic sum $p \# q$ of two real numbers p and q we understand the set of those fraction triplets which are the sum of an element of p and an element of q:

Df 1.7 $$p \# q = \iota_t(\mathrm{Re}(p) \;\&\; \mathrm{Re}(q) \;\&\; t^* \equiv \{z \mid (Ex)(Ey)(x \in p \;\&\; \&\; y \in p \;\&\; z = x +^r y)\}).$$

According to this definition we have

1.4 $$\mathrm{Re}(p) \;\&\; \mathrm{Re}(q) \rightarrow \mathrm{Re}(p \# q).$$

Further we can prove the associative and the commutative law for this arithmetic sum, as also

1.5 $$\mathrm{Re}(p) \rightarrow p \# [0] = p.$$

We also have

1.6 $$\mathrm{Re}(p) \rightarrow (Ex)(\mathrm{Re}(x) \;\&\; p \# x = [0]);$$

in fact we can prove

$$\mathrm{Re}(p) \;\&\; c^* \equiv \{z \mid \mathrm{Ftp}(z) \;\&\; (Ex)(z <^r x \;\&\; (u)(v)(w)(\langle\langle u, v\rangle, w\rangle \in p \rightarrow \rightarrow x <^r \langle\langle v, u\rangle, w\rangle))\} \cdot\!\rightarrow\!\cdot \mathrm{Re}(c) \;\&\; p \# c = [0].$$

Thus we have an additive inverse, and the unicity follows in the usual way from 1.6 together with the computation laws, so that we can define the additive inverse by

Df 1.8 $$-p = \iota_x(\mathrm{Re}(x) \;\&\; p \# x = [0]).$$

Now obviously the arithmetic difference of two real numbers p and q can be defined as the arithmetic sum of p and $-q$:

Df 1.9 $$p \ulcorner\!- q = p \# (-q),$$

and all its computation laws thus become derivable. Likewise we have

1.7 $$\mathrm{Re}(p) \rightarrow -(-p) = p \;\&\; [0] \subset p \leftrightarrow -p \subset [0].$$

The absolute value of a real number p is defined as

Df 1.10 $$|\, p \,| = \iota_t(\mathrm{Re}(t) \;\&\; [0] \subseteq t \cdot \& \cdot t = p \vee t = -p).$$

As an application we get also the concept of the distance of two real numbers p, q defining it as

Df 1.11 $$|\, p, q \,| = |\, p \ulcorner\!- q \,|.$$

We say that a real number p differs from q by less then a positive real number d if $|\, p, q \,| \subset d$.

For introducing the arithmetic product $p \,\#\, q$ of two real numbers p and q we first define an "absolute value product" $p \,\square\, q$ as the set of those fraction triplets which are either negative or the product of non negative fraction triplets belonging respectively to $|\, p \,|$ and $|\, q \,|$:

Df 1.12 $$p \,\square\, q = \iota_t(\mathrm{Re}(p) \;\&\; \mathrm{Re}(q) \;\&\; t^* \equiv \{z \,|\, \mathrm{Ftp}(z) \;\&$$
$$\& \; (z <^r \langle\langle 0, 0\rangle, 1\rangle \vee (Eu)(Ev)(u \in |\, p \,|) \;\&\; v \in |\, q \,| \;\&$$
$$\& \; \overline{u <^r \langle\langle 0, 0\rangle, 1\rangle} \;\&\; \overline{v <^r \langle\langle 0, 0\rangle, 1\rangle} \;\&\; z = u \cdot^r v)\});$$

and then we take for the arithmetic product $p \,\#\, q$ the well known definition by cases:

Df 1.13 $$p \,\#\, q = \iota_t(\mathrm{Re}(p) \;\&\; \mathrm{Re}(q) \;\&\; (([0] \subseteq p \;\&\; [0] \subseteq q) \vee$$
$$\vee (p \subseteq [0] \;\&\; q \subseteq [0]) \rightarrow t = p \,\square\, q) \;\&\; (([0] \subset p \;\&\; q \subset [0]) \vee$$
$$\vee ([0] \subset q \;\&\; p \subset [0]) \rightarrow t = -(p \,\square\, q))).$$

Here again we have

1.8 $$\mathrm{Re}(p) \;\&\; \mathrm{Re}(q) \rightarrow \mathrm{Re}(p \,\#\, q)$$

as also the associative and the commutative law. Further the distributive law between the arithmetic sum and the arithmetic product is provable. Besides we have

1.9 $$\mathrm{Re}(p) \rightarrow p \mathbin{\#\!\!\#} [1] = p.$$

Concerning the existence of a multiplicative inverse for a real number p, $p \neq [0]$ we have

1.10 $$\mathrm{Re}(p) \mathbin{\&} [0] \subset p \mathbin{\&} c^* \equiv \{z \mid \mathrm{Ftp}(z) \mathbin{\&} (Ey)(z <' y \mathbin{\&} \\ \mathbin{\&} (u)(v)(\langle\langle u, 0\rangle, v\rangle \in p \mathbin{\&} u \neq 0 \rightarrow y <' \langle\langle v, 0\rangle, u\rangle))\} \cdot\!\rightarrow\!\cdot \\ \cdot\!\rightarrow\!\cdot \mathrm{Re}(c) \mathbin{\&} p \mathbin{\#\!\!\#} c = [1],$$

and

1.11 $$\mathrm{Re}(p) \mathbin{\&} p \subset [0] \mathbin{\&} \mathrm{Re}(r) \mathbin{\&} (-p) \mathbin{\#\!\!\#} r = [1] \rightarrow p \mathbin{\#\!\!\#} (-r) = [1],$$

and again the unicity of a multiplicative inverse is provable. We therefore can define

Df 1.14 $$rc(p) = \iota_x(\mathrm{Re}(x) \mathbin{\&} p \mathbin{\#\!\!\#} x = [1]),$$

and then have

1.12 $$\mathrm{Re}(p) \mathbin{\&} p \neq [0] \rightarrow \mathrm{Re}\,(rc(p)) \mathbin{\&} p \mathbin{\#\!\!\#} rc(p) = [1].$$

Generally now the quotient $\frac{p}{q}$ of real numbers p and q, for $q \neq 0$ can be defined as $p \mathbin{\#\!\!\#} rc(q)$, with the effect that all the computation laws for the quotient become derivable.

From our definitions of the arithmetic sum and product and their inverses in particular follows the familiar isomorphism between these operations applied to rational real numbers and the corresponding operations on fractions, as for instance

1.13 $$\mathrm{Nu}(a) \mathbin{\&} \mathrm{Nu}(b) \mathbin{\&} \mathrm{Nu}(c) \mathbin{\&} \mathrm{Nu}(d) \mathbin{\&} b \neq 0 \mathbin{\&} d \neq 0 \cdot\!\rightarrow\!\cdot \\ \cdot\!\rightarrow\!\cdot \left[\frac{a}{b}\right] \# \left[\frac{c}{d}\right] = \left[\frac{a \cdot d + b \cdot c}{b \cdot d}\right] \mathbin{\&} \left(c \neq 0 \rightarrow \frac{\left[\frac{a}{b}\right]}{\left[\frac{c}{d}\right]} = \left[\frac{a \cdot d}{b \cdot c}\right]\right).$$

Finally also the laws relating to order in connection with the arithmetic operations can be seen to be satisfied. So the system of real numbers has with respect to the defined relations and operations all the properties of an ordered field, [1] being the multiplicative unit.

We still have to verify the property of continuity peculiar to the system of real numbers. Among the various formulations of the property of continuity we choose that one, especially handy for the applications, by the principle of the last upper bound:

1.14 $A \neq \Lambda$ & $(x)(x \in A \rightarrow \mathrm{Re}(x))$ & $(Ey)(\mathrm{Re}(y)$ & $(x)(x \in A \rightarrow$
$\rightarrow x \subseteq y) \cdot \twoheadrightarrow \cdot (Ez)(\mathrm{Re}(z)$ & $(u)(z \subseteq u \leftrightarrow (x)(x \in A \rightarrow x \subseteq u))$.

> «For every non void class of real numbers A which has an upper bound there exists a real number which is the least upper bound».

Indeed the sum of the elements of A, being a class of fraction triplets, is represented by a set, and this set has all the required properties. Obviously in full analogy to 1.14 we have also the theorem of the existence of a greatest lower bound for any non void class of real numbers which has a lower bound.

In the principle 1.14 the full continuity property of the system of real numbers is included. In particular it entails the archimedian character of the order of real numbers. But this one results more directly from our definition of real number. In fact the archimedian property can be formulated by the statement that for every positive real number p there exists a fraction $\frac{1}{n}$ such that $\left[\frac{1}{n}\right] \subset p$, and this is a direct consequence of the fact that every positive real number has a positive fraction triplet and therefore also a fraction $\frac{1}{n}$ as element; indeed from $\frac{1}{n} \in p$ follows $\left[\frac{1}{n}\right] \subset p$.

By this way the theory of real numbers can be established on the basis of our axioms A 1, A 2, A 3 and A 6. The passage to the n-dimensional continuum goes in the familiar way by forming pairs, triplets, and so on, of real numbers. But also the construction of the Hilbert space is possible, since we have at our disposal ω-sequences of real numbers.

As to the theory of functions we have first that a real function in the sense of analysis is a function in our sense assigning to every real number or to those of a certain class of real numbers, again a real number. Mostly the domain of such a function is an interval, that is the class of real numbers from a to b or else between a and b, where a and b are certain fixed real numbers with $a \subset b$.

For the handling with real functions it is in many cases desirable to have functions as functional sets. This obviously can be effected by using A 4. But for most of the function theory this application of A 4 is not required. Namely first continuous functions are uniquely determined by their values for rational arguments, and also stückweise continuous functions by their values for a denumerable set of arguments. Further in the theory of arbitrary functions, especially in that one of the function spaces, functions are regarded as equal if their values differ only for a set of arguments of Lebesgue measure 0. On the other hand it is known that every measurable function is – up to a set of arguments of measure 0 – the limit of an ω-sequence of scale functions belonging to a denumerable class, so that every measurable function is determined, up to a set of measure 0, by a sequence of natural numbers.

So for function theory and also for differential geometry, as also for the theory of measure and the Hilbert space we can on principle get along with the axioms A 1, A 2, A 3, A 6.

Nevertheless this procedure has its artificialities, and it seems desirable, especially for dealing with point sets, to have the continuum and its subclasses to be represented by sets. For this purpose it is not necessary to take the full potency axiom. But a sufficient measure is to take instead of the axiom of infinity the stronger "axiom of continuum" (introducing a symbol γ)

AC $$a \in \gamma \leftrightarrow (x)(x \in a \rightarrow \mathrm{Nu}(x)).$$

«γ represents the class of all number sets.»

Obviously in presence of A 4 and A 6 we immediately get AC by defining γ as $\pi(\omega)$. On the other hand from AC we can first derive VI, 5.2 by means of II, 3.8, since $\{x \mid \mathrm{Nu}(x)\}$ is in a one-to-one correspondence with $\{z \mid (Ex)(\mathrm{Nu}(x) \;\&\; z = [x])\}$, which is a subclass of γ; and hence we can define ω as the set representing $\{x \mid \mathrm{Nu}(x)\}$. Moreover we can prove by AC that the class of real numbers is represented by a set. Namely in virtue of the denumerability of

$(\omega \times \omega) \times \omega$ we have a one-to-one correspondence of the real numbers with certain subsets of ω, i.e. elements of γ, so that $\{x \mid \mathrm{Re}(x)\}$ is in a one-to-one correspondence with a subclass of γ and therefore is represented by a set.

It is further to be observed that for certain inferences in analysis the application of the axiom of choice is required. As a simple instance we take the principle often used that for every limit point of a class of real numbers C there exists an ω-sequence of elements of C converging to it. Here limit point is defined as a real number q such that for every positive real number d there exists an element of C different from q which differs from q by less than d, and an ω-sequence s is said to converge to q if for every positive real number d there exists a natural number n such that for every natural number m greater than n, $s\text{`}m$ differs from q by less than d. Obviously it is sufficient to prove the assertion for the case that $q \notin C$. We consider the class

$$\{xy \mid \mathrm{Nu}(x) \;\&\; y \in C \;\&\; |\, y, q \,| \subset \left[\frac{1}{x'}\right]\}.$$

This class, we call it Q, by our premise on q has the domain ω, and the converse domain of Q is a subclass of $\{y \mid \mathrm{Re}(y)\}$ and thus represented by a set, hence Q itself is represented by a set. Now applying to this set our axiom A 5 we can infer the existence of an ω-sequence s such that $s\text{`}n$ differs from q by less than $\left[\frac{1}{n'}\right]$ and which is therefore convergent, since for every positive real number d there exists a natural number m such that $\left[\frac{1}{m}\right] \subseteq d$.

Note that here besides the axiom of choice also AC has to be applied [1]). By this axiom the application of A 4 becomes dispensable in analysis. However, it is to be observed that, when we replace A 4 by AC, we do no more obtain, as it seems, the theorem that the continuum can be wellordered. By this way, even with adopting AC and the axiom of choice A 5, a separation between analysis and a more extended set theory is possible.

[1]) In the treatment of [Bernays 1942] AC was not required for this proof, because one there had a somewhat stronger form of the axiom of choice at disposal, referring to pairclasses instead of pair sets.

VII, § 2. Some topics of ordinal arithmetic

The fundaments of ordinal arithmetic are contained in the general theory of ordinals (III, § 1, § 2) and in the definition of the ordinal arithmetic functions $a+b, a\cdot b, a^b$ by means of the transfinite iterator (IV, § 2). We shall not enter here in all the details of ordinal arithmetic, which is not problematic from the the axiomatic point of view, but only exhibit some particular topics which have applications in cardinal arithmetic.

For this we shall employ the method of introducing specialized variables. We take small greek letters α, β, γ, δ, $\varkappa$, λ, ν, ϱ, σ for free variables and ξ, η, ζ for bound variables, ranging over the class $\{x \mid \mathrm{Od}(x)\}$. However we preserve the old bound variables in connection with the sum and the μ-symbol, since here the application of the new variables would give no advantage.

The sum operator in ordinal theory has especially the rôle of the limit operator, what is due to the following relations which hold in virtue of the fundamental properties of ordinals, in particular III, 1.17 and III, 2.3, 2.8–2.10 – (we use here the letters $\mathfrak{u}$, $\mathfrak{v}$ to denote bound greek variables):

2.1 $\quad \mathrm{Lim}(\alpha)\ \&\ (\mathfrak{u})(\mathfrak{v})(\mathfrak{u} \in \mathfrak{v}\ \&\ \mathfrak{v} \in \alpha \rightarrow \mathfrak{t}(\mathfrak{u}) \in \mathfrak{t}(\mathfrak{v})\ \&$
$\quad\quad \&\ \mathrm{Od}(\mathfrak{t}(\mathfrak{v}))\ \&\ l = \sum_{\mathfrak{x}}(\alpha, \mathfrak{t}(\mathfrak{x})) \cdot\!\rightarrow\!\cdot$

a) $\quad \cdot\!\rightarrow\!\cdot \mathrm{Lim}(l)$

b) $\quad \cdot\!\rightarrow\!\cdot \nu \in \alpha \rightarrow \mathfrak{t}(\nu) \in l$

c) $\quad \cdot\!\rightarrow\!\cdot \nu \in \alpha \rightarrow \nu \in \mathfrak{t}(\nu) \vee \nu = \mathfrak{t}(\nu)$

d) $\quad \cdot\!\rightarrow\!\cdot l = \mu_{\mathfrak{x}}(\mathfrak{u})(\mathfrak{u} \in \alpha \rightarrow \mathfrak{t}(\mathfrak{u}) \in \mathfrak{x})$

e) $\quad \cdot\!\rightarrow\!\cdot \varkappa \in l \rightarrow (E\mathfrak{u})(\mathfrak{u} \in \alpha\ \&\ \varkappa \in \mathfrak{t}(\mathfrak{u}))$.

We now first transcribe the recursive formulas IV, 2.3–2.5 and the computation laws IV, 2.6 for $a+b$, $a\cdot b$ according to the general devices for spezialized variables in I, § 1:

2.2 $$\mathrm{Od}(\alpha+\beta), \quad \mathrm{Od}(\alpha\cdot\beta)$$

2.3 a) $$\begin{cases} \alpha+0=\alpha \\ \alpha+\beta' = (\alpha+\beta)' \\ \mathrm{Lim}(\lambda) \rightarrow \alpha+\lambda = \sum_{x}(\lambda, \alpha+x), \end{cases}$$ b) $$\begin{cases} \alpha\cdot 0 = 0 \\ \alpha\cdot\beta' = \alpha\cdot\beta+\alpha \\ \mathrm{Lim}(\lambda) \rightarrow \alpha\cdot\lambda = \sum_{x}(\lambda, \alpha\cdot x) \end{cases}$$

2.4
$$\begin{cases} \alpha+(\beta+\gamma)=(\alpha+\beta)+\gamma \\ \alpha\cdot(\beta\cdot\gamma)=(\alpha\cdot\beta)\cdot\gamma \\ \alpha\cdot(\beta+\gamma)=\alpha\cdot\beta+\alpha\cdot\gamma \end{cases}$$

Besides these laws we have those which we get from the general properties of Normalfunktionen. By IV, 2.8, IV, 2.10 we have

2.5 a)
$$\begin{cases} \alpha+\beta=\alpha+\gamma \leftrightarrow \beta=\gamma \\ \alpha+\beta \in \alpha+\gamma \leftrightarrow \beta \in \gamma \end{cases}$$

2.5 b)
$$\begin{cases} 0 \in \alpha \cdot\!\rightarrow\!\cdot \alpha\cdot\beta=\alpha\cdot\gamma \leftrightarrow \beta=\gamma \\ 0 \in \alpha \cdot\!\rightarrow\!\cdot \alpha\cdot\beta \in \alpha\cdot\gamma \leftrightarrow \beta \in \gamma. \end{cases}$$

The formulas 2.5 a) and 2.5 b) come in particular to be applied for the introduction of the inverse processes of ordinal addition and multiplication: the *subtraction* of α from an ordinal β not lower than α, and the *division* of α by $\beta(\neq 0)$ with the rest δ lower than β.

The possibility and unicity of subtraction is given by the formula

2.6 $$\alpha \subseteq \beta \rightarrow (E\xi)(\alpha+\xi = \beta)$$

together with 2.5 a). The proof of 2.6 goes by considering the ordinal $\mu_x(\beta \in \alpha+x)$. This can neither be null nor a limit number, —the last owing to the last formula of 2.3 a), 2.1 e) and the property of the μ-operator—and thus there is a $\varkappa$ such that

$$\varkappa'=\mu_x(\beta \in \alpha+x);$$

hence by III, Df 1.5, III, Df 1.6

$$\alpha+\varkappa \subseteq \beta, \qquad \beta \in \alpha+\varkappa',$$

and these two formulas together with $\alpha+\varkappa'=(\alpha+\varkappa)'$ give $\alpha+\varkappa=\beta$. Note that the resulting subtraction consists only as a onesided inverse of addition.

The statement on the existence of division is

2.7 $$\beta\neq 0 \rightarrow (E\xi)(E\eta)(\eta \in \beta \;\&\; \alpha=\beta\cdot\xi+\eta).$$

The proof begins analogously as that of 2.6 with considering the ordinal $\mu_x(\alpha \in \beta\cdot x)$. Since we have

[1] $$\beta\neq 0 \rightarrow \mathrm{Suc}(\mu_x(\alpha \in \beta\cdot x)),$$

there is upon $\beta \neq 0$ a $\varkappa'$ such that

$$\beta\cdot\varkappa \subseteq \alpha, \quad \alpha \in \beta\cdot\varkappa';$$

thus by 2.6 there is a δ such that

$$\beta\cdot\varkappa+\delta = \alpha$$

and

$$\beta\cdot\varkappa+\delta \in \beta\cdot\varkappa+\beta,$$

so that by 2.5 a) we have $\delta \in \beta$.

Besides the existential statement 2.7 also the unicity formula

2.8 $\beta\neq 0 \;\&\; \gamma \in \beta \;\&\; \delta \in \beta \;\&\; \beta\cdot\varkappa+\gamma=\beta\cdot\lambda+\delta \rightarrow \varkappa=\lambda \;\&\; \gamma=\delta$

is provable as a consequence of 2.5 b) and 2.5 a).

Taking in 2.7 ω for β and using $\gamma \in \omega \rightarrow \mathrm{Nu}(\gamma)$ we get

2.9 $$(E\xi)(Ex)(\alpha=\omega\cdot\xi+x \;\&\; \mathrm{Nu}(x)),$$

which in particular entails, by III, 3.3

2.10 $$\mathrm{Lim}(\lambda) \rightarrow (E\xi)(\lambda=\omega\cdot\xi),$$

while on the other hand

2.11 $$\beta\neq 0 \rightarrow \mathrm{Lim}(\omega\cdot\beta).$$

From the stated laws on subtraction and division we now draw some consequences to be used for cardinal arithmetic.

First from the formula 2.9 we can infer that any limit number as a set of ordinals consists of *ω-sections* i.e. sets consisting of the ordinals $\alpha+n$, with $\mathrm{Nu}(n)$, for a fixed α with $\overline{\mathrm{Suc}(\alpha)}$. Indeed in virtue of the formula [1] of the proof of 2.7 we have

2.12 $$\omega\cdot\varkappa = \sum_x(\varkappa,\, \omega\cdot x' - \omega\cdot x);$$

and therefore by 2.10

2.13 $$\mathrm{Lim}(\lambda) \rightarrow (E\xi)(\lambda=\sum_x(\xi,\, \omega\cdot x' - \omega\cdot x).$$

Thus every limit number can be decomposed into mutually exclusive sets $\omega\cdot\nu' - \omega\cdot\nu$, but

$$(\omega\cdot\nu' - \omega\cdot\nu)^* \equiv \{\xi \mid (Ex)(\mathrm{Nu}(x) \;\&\; \xi=\omega\cdot\nu+x)\},$$

and thus $\omega\cdot\nu' - \omega\cdot\nu$ is an ω-section.

Another application of 2.9 is to the proof, that every infinite ordinal is of equal power with a limit number. In virtue of 2.11 it suffices for this to prove

2.14 $$\mathrm{Nu}(n)\ \&\ \varkappa \neq 0 \rightarrow \omega\cdot\varkappa + n \sim \omega\cdot\varkappa,$$

which results by showing that under our premises the class of pairs

$$\{uv \mid (Ez)(z \in n\ \&\ u = \omega\cdot\varkappa + z\ \&\ v = z) \vee u \in \omega\ \&\ v = n + u) \vee \vee\ (u \in \omega\cdot\varkappa\ \&\ u \notin \omega\ \&\ v = u))\}$$

is a one-to-one correspondence between $\omega\cdot\varkappa+n$ and $\omega\cdot\varkappa$.

Still by means of 2.7 and 2.8 we derive

2.15 $$\alpha \times \beta \sim \beta\cdot\alpha.$$

For this we first state

$$\mathrm{Crs}(\{uv \mid (E\xi)(E\eta)(u = \langle\xi, \eta\rangle\ \&\ v = \beta\cdot\xi + \eta\ \&\ \xi \in \alpha\ \&\ \eta \in \beta)\})$$

which is a consequence of 2.8. The domain of this one-to-one correspondence is obviously $\alpha \times \beta$; hence for proving 2.15 it suffices to show that the converse domain is $\beta\cdot\alpha$, formally

$$\gamma \in \beta\cdot\alpha \leftrightarrow (E\xi)(E\eta)(\gamma = \beta\cdot\xi + \eta\ \&\ \xi \in \alpha\ \&\ \eta \in \beta)$$

what follows by 2.7, 2.5 a) and 2.5 b).

A concept essential for ordinal theory is that of an *ascending ordinal limit sequence*, or briefly limit sequence. By this we understand a sequence which represents a strictly monotonic ordinal function whose domain is a limit number:

Df 2.1 $$\mathrm{Sqa}(s) \leftrightarrow \mathrm{Sq}(s)\ \&\ \mathrm{Lim}(\Delta_1 s)\ \&\ (x)(y)(u)(v)(\langle x, y\rangle \in s\ \&\ \&\ \langle u, v\rangle \in s\ \&\ x \in u \rightarrow \mathrm{Od}(v)\ \&\ y \in v).$$

There is a direct application of 2.1 a)–e) to limit sequences, since

2.16 $$\mathrm{Sqa}(s) \rightarrow \mathrm{Lim}(\Delta_1 s)\ \&\ (\xi)(\eta)(\xi \in \eta\ \&\ \eta \in \Delta_1 s \rightarrow \rightarrow s`\xi \in s`\eta\ \&\ \mathrm{Od}(s`\eta)).$$

Therefore the conclusions of 2.1 under the premise

$$\mathrm{Sqa}(s)\ \&\ \Delta_1 s = \alpha\ \&\ l = \sum_x (\alpha, s`x)$$

hold with $\mathfrak{t}(\mathfrak{a})$ replaced everywhere by $s`\mathfrak{a}$.

About composing of limit sequences we have the theorem

2.17 $$\mathrm{Sqa}(s) \mathbin{\&} \mathrm{Sqa}(t) \mathbin{\&} \nu = \Delta_1 s \mathbin{\&} \varkappa = \Delta_1 t \mathbin{\&} \lambda = \sum_x(\nu, s{`}x) \mathbin{\&}$$
$$\mathbin{\&} \nu = \sum_y(\varkappa, t{`}y) \rightarrow \lambda = \sum_y(\varkappa, s{`}t{`}y).$$

The equality to be proved under the premises, by A 3 comes back to

$$(E\xi)(\xi \in \sum_y(\varkappa, t{`}y) \mathbin{\&} \alpha \in s{`}\xi) \leftrightarrow (E\xi)(\xi \in \varkappa \mathbin{\&} \alpha \in s{`}t{`}\xi)$$

and further to

$$(E\xi)(E\eta)(\eta \in \varkappa \mathbin{\&} \xi \in t{`}\eta \mathbin{\&} \alpha \in s{`}\xi) \leftrightarrow (E\xi)(\xi \in \varkappa \mathbin{\&} \alpha \in s{`}t{`}\xi).$$

But this follows from the implications

[1] $$\beta \in \varkappa \mathbin{\&} \gamma \in t{`}\beta \mathbin{\&} \alpha \in s{`}\gamma \rightarrow \beta \in \varkappa \mathbin{\&} \alpha \in s{`}t{`}\beta,$$

which holds by the monotonity of s, and

[2] $$\beta \in \varkappa \mathbin{\&} \alpha \in s{`}t{`}\beta \rightarrow \beta' \in \varkappa \mathbin{\&} t{`}\beta \in t{`}\beta' \mathbin{\&} \alpha \in s{`}t{`}\beta$$

which results by $\mathrm{Sqa}(t)$ & $\Delta_1 t = \varkappa$.

The theorem 2.17 has an immediate application to Normalfunktionen since for every limit number ν the ν-segment of a Normalfunktion F is a limit sequence, so that we have

$$\mathrm{Nft}(F) \mathbin{\&} \mathrm{Sqa}(t) \mathbin{\&} \varkappa = \Delta_1 t \mathbin{\&} \nu = \sum_y(\varkappa, t{`}y) \rightarrow \sum_x(\nu, F{`}x) = \sum_y(\varkappa, F{`}t{`}y).$$

Since by the continuity of F we have

$$F{`}\nu = \sum_x(\nu, F{`}x)$$

we get

2.18 $$\mathrm{Nft}(F) \mathbin{\&} \mathrm{Sqa}(t) \mathbin{\&} \varkappa = \Delta_1 t \rightarrow F{`}(\sum_y(\varkappa, t{`}y)) = \sum_y(\varkappa, F{`}t{`}y).$$

By this formula a generalized continuity property of Normalfunktionen is expressed.

With the aid of 2.18 we are able to prove the existence of critical points for any Normalfunktion F, as announced in IV, § 2. We even show that there exists beyond every ordinal $\varkappa$ a critical point of F. The formal statement is

2.19 $$\mathrm{Nft}(F) \rightarrow (E\eta)(\alpha \in \eta \mathbin{\&} F{`}\eta = \eta).$$

For the proof we first observe that in the case $F`\alpha' = \alpha'$, 2.19 holds. Besides this case is only possible if F has successors as values. In the other case

$$[1] \qquad \alpha' \in F`\alpha'$$

we consider the numeral iterator $\mathsf{J}(F, \alpha')$ of F on α'. From [1] and the strict monotonity of F we get by numeral induction

$$n \in \omega \rightarrow \mathsf{J}(F, \alpha')`n \in \mathsf{J}(F, \alpha')`n'$$

and hence also

$$[2] \qquad n \in m \;\&\; m \in \omega \rightarrow \mathsf{J}(F, \alpha')`n \in \mathsf{J}(F, \alpha')`m.$$

Therefore, by 2.1 a), $\sum_x(\omega, \mathsf{J}(F, \alpha')`x)$ is a limit number, and it only remains to show

$$[3] \qquad F`\sum_x(\omega, \mathsf{J}(F, \alpha')`x) = \sum_x(\omega, \mathsf{J}(F, \alpha')`x).$$

But this follows from 2.18, using $\mathrm{Sqa}(\iota_x(x^* \equiv \mathsf{J}(F, \alpha')))$; namely substituting this ι-term for t, we first get

$$F`\sum_x(\omega, \mathsf{J}(F, \alpha')`x) = \sum_x(\omega, F`(\mathsf{J}(F, \alpha')`x))$$

and from this [3] is obtained by $\mathrm{Lim}(\omega)$ and the iteration theorem.

From 2.19 in particular follows that for every critical point of a Normalfunktion F there is a higher one and thus also a next higher one. Further we have that for any limit sequence s of critical points of F, $\sum_x(\Delta_1 s, s`x)$ is again a critical point of F, or formally

2.20 $\quad \mathrm{Nft}(F) \;\&\; \mathrm{Sqa}(s) \;\&\; \Delta_1 s = \lambda \;\&\; (\xi)(\xi \in \lambda \rightarrow F`s`\xi = s`\xi) \cdot\!\rightarrow\!\cdot$

$$\cdot\!\rightarrow\!\cdot\, F`\sum_x(\lambda, s`x) = \sum_x(\lambda, s`x).$$

Indeed upon our premise we have by 2.18

$$F`\sum_x(\lambda, s`x) = \sum_x(\lambda, F`s`x)$$
$$= \sum_x(\lambda, s`x).$$

Therefore the critical points of a Normalfunktion F are the values of again a Normalfunktion. In fact defining

Df 2.2 $\quad \mathsf{D}(F) \equiv \mathrm{I}(\{\xi\eta \mid \eta = \mu_x(\xi \in x \;\&\; F\text{'}x = x)\}, \mu_x(F\text{'}x = x))$

we have by 2.19 and IV, 2.9

2.21 $$\mathrm{Nft}(F) \to \mathrm{Nft}(\mathsf{D}(F))$$

and further

2.22 $$\mathrm{Nft}(F) \cdot\!\!\to\!\!\cdot\, \alpha \in \Delta_2(\mathsf{D}(F)) \leftrightarrow F\text{'}\alpha = \alpha.$$

The proof of 2.22 goes with using the formulas IV, 2.1 and 2.20–2.21.

A transfinite iteration of the class function D is not yielded by our transfinite iterator, but under certain assumptions it can be indirectly performed by operating with the segments of the Normalfunktion in question.

Still we introduce the notion of the *type of a limit number*. As we know, for every limit sequence s the sum $\sum_x(\Delta_1 s, s\text{'}x)$, which is the same as $\sum \Delta_2 s$, is a limit number λ, by 2.1 a). We call λ *generated by the sequence* s if $\mathrm{Sqa}(s) \;\&\; \lambda = \sum \Delta_2 s$. Every limit number λ is generated by a sequence, namely the class $\{\xi\eta \mid \xi \in \lambda \;\&\; \xi = \eta\}$ is represented by a limit sequence with the converse domain λ, and by III, 2.14 we have $\sum \lambda = \lambda$; of course λ can be generated by various sequences.

Regarding the sequences generating limit numbers λ with respect to their domain, we call λ *α-generatable* if there is a sequence with the domain α, generating λ, formally

Df 2.3 $\quad \mathrm{Gt}(\lambda, \alpha) \leftrightarrow (Ex)(\mathrm{Sqa}(x) \;\&\; \Delta_1 x = \alpha \;\&\; \lambda = \sum \Delta_2 x).$

Concerning this concept we state

2.23 $$\begin{cases} \text{a)} & \mathrm{Lim}(\lambda) \to \mathrm{Gt}(\lambda, \lambda) \\ \text{b)} & \mathrm{Gt}(\lambda, \nu) \to \nu \in \lambda \lor \nu = \lambda \\ \text{c)} & \mathrm{Gt}(\lambda, \nu) \;\&\; \mathrm{Gt}(\nu, \varkappa) \to \mathrm{Gt}(\lambda, \varkappa). \end{cases}$$

2.23 a) follows by the just made observation from III, 2.14; 2.23 b) by 2.1 c) and 2.23 c) by 2.17.

Now the type of a limit number λ is defined as the lowest ordinal α such that $\mathrm{Gt}(\lambda, \alpha)$:

Df 2.4 $$\zeta(\lambda) = \mu_x \mathrm{Gt}(\lambda, x)\ ^{1)}$$

About this function we prove

2.24 $$\mathrm{Gt}(\lambda, \nu) \rightarrow \zeta(\lambda) = \zeta(\nu).$$

First by 2.23 and Df 2.4 we directly have

$$\mathrm{Gt}(\lambda, \nu) \rightarrow \zeta(\lambda) \subseteq \zeta(\nu).$$

Thus it remains only to show that

$$\mathrm{Gt}(\lambda, \nu) \rightarrow \zeta(\nu) \subseteq \zeta(\lambda),$$

and for this it is sufficient to prove

[1] $$\mathrm{Gt}(\lambda, \nu) \,\&\, \mathrm{Gt}(\lambda, \varkappa) \rightarrow (E\xi)(\xi \subseteq \varkappa \,\&\, \mathrm{Gt}(\nu, \xi)).$$

This goes as follows. Upon our premise we have two limit sequences s and t, with the domains ν and $\varkappa$ respectively such that

$$\lambda = \sum_x (\nu, s^\iota x), \qquad \lambda = \sum_y (\varkappa, t^\iota y).$$

Let us denote by $\mathfrak{k}(\varrho)$ the term $\mu_x(x \in \nu \,\&\, t^\iota \varrho \in s^\iota x)$, then we first have upon our premise

[2] $$\varrho \in \varkappa \rightarrow \mathfrak{k}(\varrho) \in \nu \;\;\&\;\; t^\iota \varrho \in s^\iota \mathfrak{k}(\varrho).$$

Using this we further obtain

[3] $$\alpha \in \nu \rightarrow (E\xi)(\xi \in \varkappa \,\&\, \alpha \in \mathfrak{k}(\xi));$$

namely we get

$$\begin{aligned} &\alpha \in \nu \rightarrow s^\iota \alpha \in \lambda, \\ &s^\iota \alpha \in \lambda \rightarrow (E\xi)(\xi \in \varkappa \,\&\, s^\iota \alpha \in t^\iota \xi), \\ &\beta \in \varkappa \,\&\, s^\iota \alpha \in t^\iota \beta \rightarrow t^\iota \beta \in s^\iota \mathfrak{k}(\beta) \\ &\qquad\qquad\qquad\qquad \rightarrow s^\iota \alpha \in s^\iota \mathfrak{k}(\beta) \\ &\qquad\qquad\qquad\qquad \rightarrow \alpha \in \mathfrak{k}(\beta), \end{aligned}$$

and thus by the predicate calculus [3] results. [2] and [3] together yield

[4] $$\nu = \sum_y (\varkappa, \mathfrak{k}(y)).$$

---

[1]) ζ is the hebraic letter "zayin". The use of an hebraic letter here is motived by the fact, still to be shown, that $\zeta(\lambda)$ is a cardinal number.

From this we cannot infer $\mathrm{Gt}(\nu, \varkappa)$, since upon $\alpha \in \beta \in \varkappa$ we cannot prove $\mathfrak{k}(\alpha) \in \mathfrak{k}(\beta)$, but only $\mathfrak{k}(\alpha) \subseteq \mathfrak{k}(\beta)$. But a generating sequence for ν is given by the numeration of the different $\mathfrak{k}(\xi)$, $\xi \in \varkappa$ in their natural order. Now the domain of this numeration, as easily seen, is $\subseteq \varkappa$, and so we have $(E\xi)(\xi \subseteq \varkappa \;\&\; \mathrm{Gt}(\nu, \xi))$.

We call a limit number λ *reducible* or *irreducible* according as $\mathfrak{t}(\lambda) \in \lambda$ or $\mathfrak{t}(\lambda) = \lambda$ [1]). By 2.23 c, $\mathfrak{t}(\lambda)$ is always irreducible. From this we can further infer, that for every limit number λ, $\mathfrak{t}(\lambda)$ is a cardinal. Namely we have

2.25 $$\mathrm{Lim}(\lambda) \rightarrow \mathfrak{t}(\lambda) \subseteq \aleph(\lambda).$$

This indeed results from the following more general consideration. Let c be a one-to-one correspondence between an ordinal $\varkappa$ and a limit number λ, then leaving out those pairs $\langle \varrho, \sigma \rangle$ for which there exists a pair $\langle \alpha, \beta \rangle$ in c with $\alpha \in \varrho$ and $\sigma \in \beta$, we get a one-to-one correspondence d between subsets of $\varkappa$ and λ such that

[1] $$(\xi)(\eta)(\xi \in \Delta_1 d \;\&\; \eta \in \Delta_1 d \;\&\; \xi \in \eta \rightarrow d'\xi \in d'\eta)$$

[2] $$(\xi)(\xi \in \lambda \rightarrow (E\eta)(\eta \in \Delta_2 d \;\&\; \xi \in \eta)).$$

The last property of d follows since for every $\alpha \in \lambda$ the ordinal $c'\mu_z(z \in \varkappa \;\&\; \alpha \in c'z)$ must be an element of $\Delta_2 d$. Further by the natural order of $\Delta_1 d$ a numeration s of $\Delta_1 d$ is induced whose domain is an ordinal not higher than $\varkappa$ because $\Delta_1 d \subseteq \varkappa$. Then the composed set $s \mid d$ is an ascending limit sequence which generates λ and whose domain is not higher than $\varkappa$, so that $\mathfrak{t}(\lambda) \subseteq \varkappa$. Now applying this to the case $\aleph(\lambda) = \varkappa$ we obtain 2.25.

From 2.25 we infer

2.26 $$\mathrm{Lim}(\lambda) \rightarrow \mathfrak{t}(\lambda) = \aleph(\mathfrak{t}(\lambda))$$

using the formulas

$$\mathrm{Lim}(\lambda) \rightarrow \mathrm{Lim}(\mathfrak{t}(\lambda)), \quad \mathrm{Lim}(\lambda) \rightarrow \mathfrak{t}(\mathfrak{t}(\lambda)) = \mathfrak{t}(\lambda),$$

which result from 2.23, Df 2.4 and 2.24.

---

[1]) This distinction coincides with that of singular and regular limit numbers, as it occurs in the literature.

VII, § 3. Cardinal operations

The arithmetic of cardinals originates from the comparison of sets with regard to power. The fundamental facts in this respect are:

(1) the elementary properties of the relation $a \sim b$, II, § 4,
(2) ,, ,, ,, ,, ,, relations $a \preceq b, a \prec b$ V, § 1,
(3) in particular the Bernstein–Schröder theorem
$a \preceq b \ \& \ b \preceq a \rightarrow a \sim b$, V, 1.3,
(4) and the general comparability of sets $a \preceq b \vee b \preceq a$ VI, 3.2,
(5) the laws of computation and of substitutivity for the operations
$a \cup b, a \times b, a^{\underline{b}}$, II, § 4, VI, § 1,
(6) the inequality $a \prec 0''^{\underline{a}}$ VI, 1.10, VI, 1.3.

In order to pass from these statements to the theory of cardinals we have to use the concept of the cardinal $\aleph(a)$ of a set a (VI, Df 3.1), and the predicate of m being a cardinal number (VI, Df 3.3) with its fundamental properties VI, 3.5–VI, 3.10 [1]).

With the aid of these concepts we define the arithmetic operations on cardinal numbers. For this we shall have to employ some properties of the term $a \times [p]$. We note that we have

$$3.1 \quad \begin{cases} \text{a)} & a \times [p] \sim a \\ \text{b)} & p \neq q \rightarrow (a \times [p]) \cup (a \times [q]) = 0 \\ \text{c)} & a \subset b \rightarrow a \times [p] \subset b \times [p]. \end{cases}$$

We use $a \times [p]$ in order to get mutually exclusive representations of the power of a. This we do in particular for defining the cardinal sum of a and b

Df 3.1 $$a + b = \aleph((a \times [0]) \cup (b \times [0'])).$$

The cardinal product is defined by

Df 3.2 $$a \times b = \aleph(a \times b),$$

and the cardinal potency by

Df 3.3 $$a^{\underline{\underline{b}}} = \aleph(a^{\underline{b}}).$$

[1]) The application of the enumerated laws will not be expressly mentioned in this and the following section.

According to these definitions the operations of cardinal sum, product and potency are functions of arbitrary sets with cardinal numbers as values. But they yield immediately functions of cardinals, as results from the provable formula

3.2 $$a \sim c \,\&\, b \sim d \cdot\rightarrow\cdot a + b = c + d \,\&\, a \times b = c \times d \,\&\, a^{\underline{b}} = c^{\underline{d}},$$

which immediately gives

3.3 $$\begin{cases} \text{a)} & a + b = \aleph(a) + \aleph(b) \\ \text{b)} & a \times b = \aleph(a) \times \aleph(b) \\ \text{c)} & a^{\underline{b}} = \aleph(a)^{\underline{\aleph}(b)}. \end{cases}$$

For the sum we also note

3.4 $$a \cap b = 0 \rightarrow a + b = \aleph(a \cup b), \quad a + 0 = \aleph(a).$$

There is no difficulty to extend our definitions of cardinal sum and product to the case of arbitrary many members, namely by defining

Df 3.4 $$\sum_{\mathfrak{x}, m} \mathfrak{t}(\mathfrak{x}) = \aleph(\sum_{\mathfrak{x}}(m, \mathfrak{t}(\mathfrak{x}) \times [\mathfrak{x}]))$$

Df 3.5 $$\prod_{\mathfrak{x}, m} \mathfrak{t}(\mathfrak{x}) = \aleph(\prod_{\mathfrak{x}}(m, \mathfrak{t}(\mathfrak{x}))).$$

Corresponding to 3.2 we have here to prove [1])

3.5 $$(x)(x \in m \rightarrow \mathfrak{s}(x) \sim \mathfrak{t}(x)) \rightarrow \sum_{x, m} \mathfrak{s}(x) = \sum_{x, m} \mathfrak{t}(x)$$

3.6 $$(x)(x \in m \rightarrow \mathfrak{s}(x) \sim \mathfrak{t}(x)) \rightarrow \prod_{x, m} \mathfrak{s}(x) = \prod_{x, m} \mathfrak{t}(x).$$

For the proof of 3.5 we have to set up a one-to-one correspondence

[1]) It may be allowed to take in the following formula schemata of this section (up to p. 179) instead of the syntactical variables $\mathfrak{x}$, $\mathfrak{y}$, $\mathfrak{u}$, $\mathfrak{v}$, the corresponding latin variables. Note that we could deal here everywhere with formulas instead of schemata by applying instead of denotations of terms with arguments rather variables for functional sets; cf. II, 1.15 and II, 3.10.

between the two sums on the right. To this end we consider the class of pairs

[1] $\{xy \mid x \in m \,\&\, \mathrm{Crs}(y) \,\&\, \Delta_1 y = \mathfrak{s}(x) \times [x] \,\&\, \Delta_2 y = \mathfrak{t}(x) \times [x]\}$.

The domain of [1] is m, since by the premise of 3.5 there exists for every x out of m a one-to-one correspondence between $\mathfrak{s}(x) \times [x]$ and $\mathfrak{t}(x) \times [x]$. In order to apply to [1] the axiom of choice A 5 we have to show that this class is represented by a set. This follows by II, 3.14 from the fact that the converse domain of [1] is a subclass of $\sum_x(m, \pi((\mathfrak{s}(x) \times [x]) \times (\mathfrak{t}(x) \times [x])))$. Now as a consequence of A 5 there exists a functional set f assigning to every element x of m a one-to-one correspondence between $\mathfrak{s}(x) \times [x]$ and $\mathfrak{t}(x) \times [x]$, moreover $\sum_x(m, f\text{'}x)$ is a one-to-one correspondence between $\sum_x(m, \mathfrak{s}(x) \times [x])$ and $\sum_x(m, \mathfrak{t}(x) \times [x])$, so that 3.5 results.

The proof of 3.6 goes by first showing, with applying A 5 in a corresponding way as above, the existence of a function g assigning to every element x of m a one-to-one correspondence between $\mathfrak{s}(x)$ and $\mathfrak{t}(x)$. Now it results that the class of pairs

$$[2] \qquad \{uv \mid u \in \prod_x(m, \mathfrak{s}(x)) \,\&\, \mathrm{Ft}(v) \,\&\, \Delta_1 v = m \,\&\, (x)(x \in m \to \\ \to \langle u\text{'}x, v\text{'}x\rangle \in g\text{'}x)\}$$

is a one-to-one correspondence between the two products $\prod_x(m, \mathfrak{s}(x))$ and $\prod_x(m, \mathfrak{t}(x))$; at the same time it follows that [2] is represented by a set.

Parallel to the formula 3.4 we also have

$$3.7 \quad (x)(y)(x \in m \,\&\, y \in m \,\&\, x \neq y \to \mathfrak{s}(x) \cap \mathfrak{s}(y) = 0) \to \\ \to \sum_{x,\, m} \mathfrak{s}(x) = \aleph(\sum_x(m, \mathfrak{s}(x))).$$

Indeed a one-to-one correspondence between $\sum_x(m, \mathfrak{s}(x) \times [x])$ and $\sum_x(m, \mathfrak{s}(x))$ is given by the class

$$\{uv \mid (Ex)(x \in m \,\&\, v \in \mathfrak{s}(x) \,\&\, u = \langle v, x\rangle)\}.$$

An advantage of our way of introducing cardinal numbers is that we need not with respect to them a particular definition of equality and of smaller than, since equality of cardinal numbers is the same as set equality and the relation smaller than between cardinal numbers is the same as the proper subset relation as also the element relation between ordinals. Further all the elementary computation laws hold unrestrictedly for the cardinal operations in virtue of the corresponding laws for the defining operations; thus for instance we have

$$a \times (b + c) = (a \times b) + (a \times c).$$

We also get

3.8
$$\begin{cases} a \preceq b \rightarrow a + c \subseteq b + c \\ \qquad\quad \rightarrow a \times c \subseteq b \times c \\ \qquad\quad \rightarrow a^{\underline{c}} \subseteq b^{\underline{c}} \\ a \preceq b \;\&\; c \neq 0 \rightarrow c^{\underline{a}} \subseteq c^{\underline{b}} \end{cases}$$

and correspondingly for infinite cardinal sums and products, by 3.5 and 3.6,

3.9
$$(x)(x \in m \rightarrow \mathfrak{s}(x) \preceq \mathfrak{t}(x)) \cdot\!\rightarrow\!\cdot \sum_{x,\,m} \mathfrak{s}(x) \subseteq \sum_{x,\,m} \mathfrak{t}(x) \;\&\; \prod_{x,\,m} \mathfrak{s}(x) \subseteq \prod_{x,\,m} \mathfrak{t}(x).$$

We also note the formula

3.10
$$a \neq 0 \rightarrow \aleph(b) \subseteq a \times b.$$

The analogy with ordinary arithematic appears likewise in the following equations concerning sums and products with equal members

3.11
$$\sum_{x,\,m} \aleph(a \times [x]) = m \times \aleph(a)$$

3.12
$$\prod_{x,\,m} \aleph(a \times [x]) = \aleph(a)^{\underline{m}}.$$

The first results from the formula

$$\sum_x (m, a \times [x]) = a \times m$$

together with 3.5; the second follows from VI, 1.5 together with 3.6.

From 3.11 and 3.12 in connection with 3.9 we get

3.13 $(x)(x \in m \rightarrow \mathfrak{s}(x) \preceq a) \cdot \rightarrow \cdot \sum_{x, m} \mathfrak{s}(x) \subseteq m \times a \;\&\; \prod_{x, m} \mathfrak{s}(x) \subseteq a^{\underline{m}}.$

Also the following distributive laws are valable

3.14 $$\Big(\sum_{x, m} \mathfrak{t}(x)\Big) \times a = \sum_{x, m} (\mathfrak{t}(x) \times a)$$

3.15 $$\Big(\prod_{x, m} \mathfrak{t}(x)\Big)^{\underline{a}} = \prod_{x, m} (\mathfrak{t}(x)^{\underline{a}}).$$

The proof of 3.14 goes by starting from the identity

$$\sum_x (m, \mathfrak{t}(x)) \times a = \sum_x (m, (\mathfrak{t}(x) \times a))$$

which then has to be applied with $\mathfrak{t}(x) \times [x]$ instead of $\mathfrak{t}(x)$ and using the formula 3.5 and

$$(c \times [d]) \times a \sim (c \times a) \times [d].$$

For proving 3.15 the one-to-one correspondence to be set up consists in coordinating with each function u which assigns to every y, $y \in a$, a function assigning to every x, $x \in m$, a value $(u\text{'}y)\text{'}x$, out of $\mathfrak{t}(x)$, that function v which assigns to every x, $x \in m$, the function assigning to every y, $y \in a$, the value $(u\text{'}y)\text{'}x$. Formally the one-to-one correspondence is

$$\{uv \mid \mathrm{Ft}(u) \;\&\; \Delta_1 u = a \;\&\; (y)(y \in a \rightarrow \mathrm{Ft}(u\text{'}y) \;\&\; \Delta_1(u\text{'}y) = m \;\&$$
$$\& \; (x)(x \in m \rightarrow (u\text{'}y)\text{'}x \in \mathfrak{t}(x))) \;\&$$
$$\& \; \mathrm{Ft}(v) \;\&\; \Delta_1 v = m \;\&\; (x)(x \in m \rightarrow \mathrm{Ft}(v\text{'}x) \;\&\; \Delta_1(v\text{'}x) = a \;\&$$
$$\& \; (y)(y \in a \rightarrow (v\text{'}x)\text{'}y = (u\text{'}y)\text{'}x))\}.$$

The laws for the relation smaller than (with exclusion of equality) are most not generally valable for infinite sets. But at all events we have the theorem of J. König, which in its generalized form, given by Ph. Jourdain, says

3.16 $(x)(x \in m \rightarrow \mathfrak{s}(x) \prec \mathfrak{t}(x)) \cdot \rightarrow \cdot \sum_{x, m} \mathfrak{s}(x) \subset \prod_{x, m} \mathfrak{t}(x).$

For the proof we first note that, as a special application of 3.6, we have

[1] $$\prod_x(m, \mathfrak{t}(x)) \sim \prod_x(m, \mathfrak{t}(x) \times [x]),$$

so that 3.16 will follow if we prove

[2] $$(x)(x \in m \rightarrow \mathfrak{s}(x) \prec \mathfrak{t}(x) \cdot\rightarrow\cdot \sum_x(m, \mathfrak{s}(x) \times [x]) \prec \prod_x(m, \mathfrak{t}(x) \times [x]);$$

moreover we have

$$(x)(x \in m \ \& \ \mathfrak{s}(x) \prec \mathfrak{t}(x))\cdot\rightarrow\cdot\mathfrak{s}(x) \times [x] \prec \mathfrak{t}(x) \times [x].$$

and

$$(x)(y)(x \neq y \rightarrow (\mathfrak{s}(x) \times [x]) \cap (\mathfrak{s}(y) \times [y]) = 0),$$

and the same for $\mathfrak{t}$.

Thus for obtaining [2] it is sufficient to prove

[3] $$(x)(x \in m\cdot\rightarrow\cdot\mathfrak{k}(x) \prec \mathfrak{l}(x) \ \& \ (x)(y)(x \neq y \rightarrow \mathfrak{k}(x) \cap \mathfrak{k}(y) = 0 \ \& \ \& \ \mathfrak{l}(x) \cap \mathfrak{l}(y) = 0) \ :\rightarrow: \ \sum_x(m, \mathfrak{k}(x)) \prec \prod_x(m, \mathfrak{l}(x)).$$

This is to be done by

(i) setting up a one-to-one correspondence between $\sum_x(m, \mathfrak{k}(x))$ and a subset of $\prod_x(m, \mathfrak{l}(x))$

(ii) proving that for every mapping of $\sum_x(m, \mathfrak{k}(x))$ into $\prod_x(m, \mathfrak{l}(x))$ by a function f there is an element of $\prod_x(m, \mathfrak{l}(x))$ which is not in the converse domain of f.

For (i) we first infer from

$$(x)(x \in m \rightarrow \mathfrak{l}(x) \bar{\ } \mathfrak{k}(x) \neq 0),$$

which consists under our premise by V, 1.7, with the aid of the axiom of choice in the form VI, 2.2, the existence of a functional set g such that

$$(x)(x \in m \rightarrow g\text{`}x \in (\mathfrak{l}(x) \bar{\ } \mathfrak{k}(x)).$$

Then the one-to-one correspondence is given by the class of pairs

$$\{uv \,|\, (Ex)(x \in m \ \& \ u \in \mathfrak{k}(x) \ \& \ \mathrm{Ft}(v) \ \& \ \Delta_1 v = m \ \& \ v\text{`}x = u \ \& \ \& \ (z)(z \in m \ \& \ z \neq x \ \rightarrow \ v\text{`}z = g\text{`}z))\}$$

which is easily shown to be one-to-one.

For (ii) we have the premise on f:

[4] $$\mathrm{Ft}(f) \mathbin{\&} \Delta_1 f = \sum_x(m, \mathfrak{k}(x)) \mathbin{\&} \Delta_2 f \subseteq \prod_x(m, \mathfrak{l}(x))$$

at our disposal. From [4] follows

$$a \in \mathfrak{k}(c) \mathbin{\&} c \in m \cdot\!\!\rightarrow\!\!\cdot \mathrm{Ft}(f`a) \mathbin{\&} (b \in m \rightarrow (f`a)`b \in \mathfrak{l}(b));$$

further from the consequence VI, 2.9 of A 5 and our premise we infer that for every element c of m the functional set with the domain $\mathfrak{k}(c)$ assigning to every element of a of $\mathfrak{k}(c)$ the value $((f`a)`c$ has a converse domain of lower power than $\mathfrak{l}(c)$; therefore again by V, 1.7

$$c \in m \rightarrow (Ez)(z \in \mathfrak{l}(c) \mathbin{\&} (u)(u \in \mathfrak{k}(c) \rightarrow (f`u)`c \neq z))$$

and so the class

$$\{xz \mid x \in m \mathbin{\&} z \in \mathfrak{l}(x) \mathbin{\&} (u)(u \in \mathfrak{k}(x) \rightarrow (f`u)`x \neq z)\}$$

has the domain m. Further this class is represented by a set, since its converse domain is a subclass of $\sum_x(m, \mathfrak{l}(x))$. Therefore applying A 5 we get

$$(Ey)(y \in \prod_x(m, \mathfrak{l}(x)) \mathbin{\&} (x)(u)(x \in m \mathbin{\&} u \in \mathfrak{k}(x) \rightarrow (f`u)`x \neq y`x))$$

and from this with the predicate calculus we obtain

$$(Ey)(y \in \prod_x(m, \mathfrak{l}(x)) \mathbin{\&} (u)(u \in \sum_x(m, \mathfrak{k}(x)) \rightarrow f`u \neq y)).$$

But just this was to be proved.

VII, § 4. Formal laws on cardinals

Besides those laws of cardinal arithmetic which are common for finite and infinite cardinals, there are also some relations peculiar to transfinite arithmetic. In particular we have here that the sum and the product of two cardinals, one at least of which is infinite, equals the maximal one. For proving this it is the main point to show that for every infinite set a the cardinal sum $a + a$ and the product $a \times a$ is simply $\aleph(a)$. We prove first

4.1 $$\mathrm{Cd}(\alpha) \mathbin{\&} \omega \subseteq \alpha \rightarrow \alpha + \alpha = \alpha$$

For this we use our theorems concerning the division of ordinals with ω. From 2.9, and 2.14 we get by Df VI, 3.3 of Cd:

4.2 $$\mathrm{Cd}(\alpha) \mathbin{\&} \omega \subseteq \alpha \rightarrow (E\xi)(\xi \neq 0 \mathbin{\&} \alpha = \omega \cdot \xi).$$

Hence our proof comes out to show

$$\omega \cdot \varkappa + \omega \cdot \varkappa \sim \omega \cdot \varkappa$$

i.e. by Df 3.1

[1] $$\omega \cdot \varkappa \times [0] \cup \omega \cdot \varkappa \times [0'] \sim \omega \cdot \varkappa$$

However this results with the aid of 2.12. Namely by this we have

$$(\omega \cdot \varkappa \times [0]) \cup (\omega \cdot \varkappa \times [0']) =$$
$$= (\sum_x (\varkappa, \omega \cdot x' {}^{-} \omega \cdot x) \times [0]) \cup (\sum_x (\varkappa, \omega \cdot x' {}^{-} \omega \cdot x) \times [0']) =$$
$$= \sum_x (\varkappa, ((\omega \cdot x' {}^{-} \omega \cdot x) \times [0]) \cup ((\omega \cdot x' {}^{-} \omega \cdot x) \times [0'])).$$

Since the sets $\omega \cdot x' {}^{-} \omega \cdot x$ corresponding to different x out of $\varkappa$ are mutually exclusive the proof of [1] reduces to that of

$$(\omega \cdot \beta' {}^{-} \omega \cdot \beta) \times [0] \cup (\omega \cdot \beta' {}^{-} \omega \cdot \beta) \times [0'] \sim \omega \cdot \beta' {}^{-} \omega \cdot \beta.$$

But this directly results by verifying that the class of pairs

$$\{uv \,\|\, (Ez)(\mathrm{Nu}(z) : \& : u = \langle \omega \cdot \beta + z, 0 \rangle \mathbin{\&} v = \omega \cdot \beta + z + z \cdot \mathbf{v} \cdot$$
$$\cdot \mathbf{v} \cdot u = \langle \omega \cdot \beta + z, 0' \rangle \mathbin{\&} v = \omega \cdot \beta + z + z')\}$$

is a one-to-one correspondence.

From 4.1 we immediately get

4.3 $$\mathrm{Infin}(a) \rightarrow a + a = \aleph(a).$$

More generally from 4.3 we infer

4.4 $$a \preceq b \mathbin{\&} \mathrm{Infin}(b) \rightarrow a + b = \aleph(b).$$

Namely upon $a \preceq b$ we have

$$(a \times [0]) \cup (b \times [0']) \preceq (b \times [0]) \cup (b \times [0'])$$

hence

$$a + b \subseteq b + b$$

and by 4.3

[1] $$a + b \subseteq \aleph(b).$$

On the other hand we obviously have

[2] $$\aleph(b) \subseteq a + b.$$

From [1] and [2] together 4.4 results.

At the same time we get

4.4 a $$a \preceq b \ \& \ \mathrm{Infin}(b) \rightarrow \aleph(a \cup b) = \aleph(b).$$

Parallel to 4.1 we prove for the cardinal product

4.5 $$\mathrm{Cd}(\alpha) \ \& \ \omega \subseteq \alpha \ \rightarrow \ \alpha \times \alpha = \alpha.$$

This goes by transfinite induction on α. Thus let α be an infinite cardinal, such that for lower infinite cardinals 4.5 holds. We decompose $\alpha \times \alpha$ into three sets a, b, c representing the classes

$$a^* \equiv \{\xi\eta \mid \zeta = \eta \ \& \ \eta \in \alpha\}$$
$$b^* \equiv \{\xi\eta \mid \xi \in \eta \ \& \ \eta \in \alpha\}$$
$$c^* \equiv \{\xi\eta \mid \eta \in \xi \ \& \ \xi \in \alpha\}.$$

Since $a \sim \alpha$ and $b \sim c$, 4.5 will follow with the aid of 4.1, when we show that $c \sim \alpha$. For this we consider the lexicographic order of c, which is a wellorder and thus generates a numeration of c. Let δ be the domain of this numeration then $c \sim \delta$. Further $\delta \succeq \alpha$, and so we only still need to show that $\delta \preceq \alpha$. We even prove $\delta \subseteq \alpha$, what by $\mathrm{Cd}(\alpha)$ amounts to show that

$$\gamma \in \delta \rightarrow \aleph(\gamma) \in \alpha.$$

This now goes as follows. By the said numeration of c, γ is assigned to a pair $\langle \varkappa, \lambda \rangle$ with $\lambda \in \varkappa \ \& \ \varkappa \in \alpha$; at the same time γ is in a one-to-one correspondence with the set of pairs preceding $\langle \varkappa, \lambda \rangle$ in the lexicographic order. This set is a subset of $\varkappa' \times \varkappa'$. Here $\varkappa$ is lower than α and therefore $\aleph(\varkappa') \in \alpha$. Now either $\varkappa$ is finite, then $\varkappa' \times \varkappa'$ also is finite and therefore γ is finite. If $\varkappa$ is infinite then by the assumption of our induction $\aleph(\varkappa') \times \aleph(\varkappa') = \aleph(\varkappa')$ and hence $\aleph(\gamma) \subseteq \aleph(\varkappa' \times \varkappa') = \aleph(\varkappa')$, thus $\aleph(\gamma) \in \alpha$.

Corresponding to the passage from 4.1 to 4.4 we derive from 4.5

4.6 $$a \neq 0 \ \& \ a \preceq b \ \& \ \mathrm{Infin}(b) \ \rightarrow \ a \times b = \aleph(b),$$

using that $a \preceq b \rightarrow a \times b \preceq b \times b$ and $a \neq 0 \rightarrow b \preceq a \times b$.

As a consequence of 4.6 we get (writing the symbol 2 for $0''$):

4.7 $$\text{Infin}(a) \rightarrow 2^{\underline{a}} = a^{\underline{a}}.$$

Indeed we have

$$a \underline{\prec} [0, 0']^{\underline{a}}$$

and therefore by VI, 1.9, VI, 1.6 and 4.6

$$a^{\underline{a}} \underline{\prec} 2^{\underline{a \times a}} \sim 2^{\underline{a}}.$$

Likewise we have

4.8 $$2 \underline{\prec} c \underline{\prec} a \;\&\; \text{Infin}(a) \rightarrow 2^{\underline{a}} = c^{\underline{a}} = a^{\underline{a}}.$$

From the theorems 4.1 and 4.5 it results that for infinite cardinals the operations of addition and multiplication do not lead from two given cardinals to a higher one. On the other hand we know that $2^{\underline{a}}$ is a higher cardinal than $\aleph(a)$, so that there is for every cardinal a higher one, and thus also a next higher one. Moreover for every limit sequence s of cardinals the sum of its members is the next higher cardinal. Namely first this sum λ is by III, 2.9, 2.10 the least ordinal higher than these members. But λ must also be a cardinal since otherwise there would be a cardinal $s\text{'}\varkappa$, $\varkappa \in \Delta_1 s$, such that $\aleph(\lambda) \prec s\text{'}\varkappa \underline{\prec} \lambda$, what indeed is impossible.

By this way the sequential class $\aleph$ of infinite cardinals in their natural order is obtainable by the following transfinite iteration

Df 4.1 $$\aleph \equiv \text{I}(\{\xi\eta \mid \eta = \mu_x(\text{Cd}(x) \;\&\; \xi \in x)\}, \omega); \qquad \aleph_a = \aleph\text{'}a.$$

According to Df 4.1 $\aleph$ is a Normalfunktion, and we have $\aleph_0 = \omega$, $\aleph_{\alpha'}$ is the cardinal following immediately $\aleph_\alpha$, and

$$\text{Lim}(\nu) \rightarrow \aleph_\nu = \sum_x(\nu, \aleph_x).$$

$\aleph_{\alpha'}$ is also determined as being the cardinal number of the set representing the Zahlenklasse with the initial number $\aleph_\alpha$. Indeed this set is $\aleph_{\alpha'} {}^{-} \aleph_\alpha$ and we have

$$\aleph_{\alpha'} = \aleph_\alpha \cup (\aleph_{\alpha'} {}^{-} \aleph_\alpha), \quad \aleph_\alpha \in \aleph_{\alpha'}$$

hence by 4.4 a

$$\aleph(\aleph_{\alpha'} \,\bar{}\, \aleph_\alpha) = \aleph_{\alpha'}.$$

To the function $\aleph$ the general considerations on Normalfunktionen are applying. We have first

4.9 $$\nu \subseteq \aleph_\nu$$

and a fortiori

4.10 $$\aleph(\nu) \subseteq \aleph_\nu.$$

Further there exist the critical points of $\aleph$ which are the values of the function $\mathsf{D}(\aleph)$. We define

Df 4.2 $$\beth = \mathsf{D}(\aleph), \quad \beth_a = \beth{}^{\iota}a.$$

$\beth$ is again a Normalfunktion and its values are the cardinals ν such that

$$\aleph_\nu = \nu;$$

this equation expresses that there are as much cardinals below $\aleph_\nu$ as there are ordinals below it. Besides by our former statements on critical points 2.19, using also that every value of $\aleph$ is a limit number, we have

4.11 $$\begin{cases} \beth_0 = \sum\limits_x(\omega, \mathsf{J}(\aleph, 0)^{\iota}x) \\ \beth_{\nu'} = \sum\limits_x(\omega, \mathsf{J}(\aleph, \beth_\nu')^{\iota}x) \\ \mathrm{Lim}(\nu) \rightarrow \beth_\nu = \sum\limits_x(\nu, \beth_x). \end{cases}$$

Considering now the cardinals $\aleph_\nu$ with respect to their type, we have the following facts

4.12 $$\begin{cases} \text{ז}(\aleph_0) = \omega \\ \text{ז}(\aleph_{\nu'}) = \aleph_{\nu'}. \end{cases}$$

The first is obvious; the proof of the second one is indirect. If there were for some $\lambda \in \aleph_{\nu'}$ a limit sequence s, with $\Delta_1 s = \lambda$, such that

$$\aleph_{\nu'} = \sum_x(\lambda, s^{\iota}x),$$

we should have

$$\alpha \in \lambda \to \aleph(s^{\iota}\alpha) \subseteq \aleph_\nu ,$$

hence by Df 3.4, 3.13 and 4.6

$$\aleph(\sum_x(\lambda, s^{\iota}x) \subseteq \sum_{x,\lambda} \aleph(s^{\iota}x) \subseteq \aleph(\lambda) \times \aleph_\nu = \aleph_\nu .$$

Thus for $\nu = 0$ and Suc(ν) the cardinal $\aleph_\nu$ is irreducible. Now the question remains of the types of $\aleph_\nu$ for the case Lim(ν). Here we have by the continuity of $\boldsymbol{\aleph}$, Gt($\aleph_\nu, \nu$) and hence by 2.24

4.13 $$\mathrm{Lim}(\nu) \to \text{ז}(\aleph_\nu) = \text{ז}(\nu).$$

Therefore a necessary condition for $\aleph_\nu$, with Lim(ν), being irreducible is that $\aleph_\nu = \nu$, i.e. $\aleph_\nu$ must be a critical point of $\boldsymbol{\aleph}$, thus a value of $\boldsymbol{\beth}$. Again if an irreducible $\aleph_\nu$ with Lim(ν) is an $\beth_\varkappa$, then $\varkappa$ cannot be 0 neither a successor, since then by 4.11 $\text{ז}(\beth_\varkappa) = \omega$. Therefore $\varkappa$ must be a limit number, and now anew it results that $\beth_\varkappa$ must be a critical point of $\boldsymbol{\beth}$. In the same way the argument goes further from $\boldsymbol{\beth}$ to D($\boldsymbol{\beth}$), and so on, as was pointed to by Hausdorff. Inquiring here further one comes to state that it is impossible to infer from our axioms the existence of an irreducible cardinal $\aleph_\nu$ with limit index ν [1]).

With the aid of the function $\boldsymbol{\aleph}$ the laws of cardinal arithmetic can be formulated in a pregnant way. By 4.4 and 4.6 we have

4.14 $$\alpha \subseteq \beta \to \aleph_\alpha + \aleph_\beta = \aleph_\alpha \times \aleph_\beta = \aleph_\beta ,$$

what in particular entails

4.15 $$\alpha \subset \beta \;\&\; \gamma \subset \delta \to \aleph_\alpha + \aleph_\gamma \subset \aleph_\beta + \aleph_\delta$$
$$\to \aleph_\alpha \times \aleph_\gamma \subset \aleph_\beta \times \aleph_\delta .$$

Further we have by 4.7, 4.8

4.16 $$\aleph_\alpha \subset 2^{\underline{\aleph_\alpha}}$$

4.17 $$\alpha \subseteq \beta \to 2^{\underline{\aleph_\beta}} = \aleph_\alpha^{\underline{\aleph_\beta}} = \aleph_\beta^{\underline{\aleph_\beta}} .$$

---

1) The proof of this impossibility belongs to the questions of independence of which we intend to treat in the second volume.

The last relation comes to be applied in the proof of Hausdorff's recurrence formula [1904]

4.18 $$\aleph_{\alpha'}^{\underline{\aleph_\beta}} = \aleph_{\alpha'} \times \aleph_{\alpha}^{\underline{\aleph_\beta}}.$$

For proving this equality we first observe that generally

$$\aleph_{\alpha}^{\underline{\aleph_\beta}} \subseteq \aleph_{\alpha'}^{\underline{\aleph_\beta}} \quad , \quad \aleph_{\alpha'} \subseteq \aleph_{\alpha'}^{\underline{\aleph_\beta}}$$

and thus by 4.14

$$\aleph_{\alpha}^{\underline{\aleph_\beta}} \times \aleph_{\alpha'} \subseteq \aleph_{\alpha'}^{\underline{\aleph_\beta}}.$$

Therefore it is sufficient to show that

$$\aleph_{\alpha'}^{\underline{\aleph_\beta}} \subseteq \aleph_{\alpha}^{\underline{\aleph_\beta}} \times \aleph_{\alpha'}.$$

For this we distinguish the two cases $\alpha' \subseteq \beta$ and $\beta \subseteq \alpha$. In the first case we have by 4.17

$$\aleph_{\alpha'}^{\underline{\aleph_\beta}} = \aleph_{\alpha}^{\underline{\aleph_\beta}} \subseteq \aleph_{\alpha}^{\underline{\aleph_\beta}} \times \aleph_{\alpha'}.$$

For the second case we argue as follows. Every element of $\aleph_{\alpha'}^{\underline{\aleph_\beta}}$ is a sequence s of elements of $\aleph_{\alpha'}$ with the domain $\aleph_\beta$. If $\lambda = \mu_x(\xi)(\xi \in \Delta_2 s \rightarrow \xi \in x)$, that is the lowest ordinal higher than every value of s, then $s \in \lambda^{\underline{\aleph_\beta}}$ and $\lambda \subseteq \aleph_{\alpha'}$. But here the equality cannot hold. For otherwise the numeration of $\Delta_2 s$ by the natural order would be an ascending limit sequence t generating $\aleph_{\alpha'}$; and since $\aleph_{\alpha'}$ is an irreducible limit number, $\Delta_1 t$ would have to be $\aleph_{\alpha'}$. However this is impossible since $\Delta_1 t \preceq \Delta_1 s$, $\Delta_1 s = \aleph_\beta$, $\aleph_\beta \preceq \aleph_\alpha$. Hence $\lambda \in \aleph_{\alpha'}$. This means that the members of a sequence belonging to $\aleph_{\alpha'}^{\underline{\aleph_\beta}}$ are elements of some element of $\aleph_{\alpha'}$.

Hence we have

$$\aleph_{\alpha'}^{\underline{\aleph_\beta}} \subseteq \sum_x (\aleph_{\alpha'}, x^{\underline{\aleph_\beta}})$$

and thus

[1] $$\aleph_{\alpha'}^{\underline{\aleph_\beta}} \subseteq \sum_{x,\, \aleph_{\alpha'}} x^{\underline{\aleph_\beta}}.$$

Besides

$$c \in \aleph_{\alpha'} \rightarrow c^{\underline{\aleph_\beta}} \subseteq \aleph_{\alpha}^{\underline{\aleph_\beta}}.$$

Therefore by 3.13

[2] $$\sum_{x,\,\aleph_{\alpha'}} x^{\underline{\aleph_\beta}} \subseteq \aleph_{\alpha'} \times \aleph_{\alpha}^{\underline{\aleph_\beta}}$$

and by [1] and [2]

$$\aleph_{\alpha'}^{\underline{\aleph_\beta}} \subseteq \aleph_{\alpha'} \times \aleph_{\alpha}^{\underline{\aleph_\beta}}.$$

By the Hausdorff recurrence formula 4.18 the potency $a^{\underline{b}}$ for the case of $\aleph(a) = \aleph_{\alpha'}$ is reduced to the potency $\aleph_{\alpha}^{\underline{b}}$. For the case that $\aleph(a) = \aleph_\lambda$ with $\mathrm{Lim}(\lambda)$ a reduction formula has been given by Tarski, which however includes infinite sums and products.

For infinite sums and products there are also general evaluation laws. Concerning sums we have

4.19 $$\begin{cases} \text{a)} & \displaystyle\sum_{x,\,\lambda'} \aleph_x = \aleph_\lambda \\ \text{b)} & \mathrm{Lim}(\lambda) \rightarrow \displaystyle\sum_{x,\,\lambda} \aleph_x = \aleph_\lambda . \end{cases}$$

4.19 a) follows by 3.13, 4.10 and 4.14. Indeed we get

$$\aleph_\lambda \subseteq \sum_{x,\,\lambda'} \aleph_x \subseteq \aleph(\lambda) \times \aleph_\lambda \subseteq \aleph_\lambda \times \aleph_\lambda \subseteq \aleph_\lambda .$$

At the same time we get

[1] $$\sum_{x,\,\lambda} \aleph_x \subseteq \aleph_\lambda .$$

Besides for a limit number λ we have

$$\aleph_\lambda = \sum_x (\lambda, \aleph_x)$$

and hence

[2] $$\aleph_\lambda \subseteq \sum_{x,\,\lambda} \aleph_x .$$

From [1] and [2] results 4.19 b).

In analogy to the formulas 4.19 the following formulas for infinite products hold

$$
4.20 \quad \begin{cases} \text{a)} & \prod\limits_{x,\lambda'} \aleph_x = \aleph_\lambda^{\underline{\aleph(\lambda)}} \\ \text{b)} & \mathrm{Lim}(\lambda) \rightarrow \prod\limits_{x,\lambda} \aleph_x = \aleph_\lambda^{\underline{\aleph(\lambda)}} \end{cases}
$$

The proof of these formulas, which was given by Tarski [1925], requires somewhat more entering in the theory of ordinals. We refrain here from carrying it out [1]).

The formulas 4.19 b and 4.20 b can in particular be used for proving the mentioned reduction laws of Tarski for potencies $\aleph_\alpha^{\underline{\aleph_\beta}}$ with $\mathrm{Lim}(\alpha)$:

$$
4.21 \quad \begin{cases} \text{a)} & \mathrm{Lim}(\alpha) \mathbin{\&} \aleph_\beta \subset \daleth(\alpha) \;\cdot\!\!\rightarrow\!\!\cdot\; \aleph_\alpha^{\underline{\aleph_\beta}} = \sum\limits_{x,\alpha} \aleph_x^{\underline{\aleph_\beta}} \\ \text{b)} & \mathrm{Lim}(\alpha) \mathbin{\&} \daleth(\alpha) \subseteq \aleph_\beta \mathbin{\&} \mathrm{Sqa}(s) \mathbin{\&} \sum\limits_x (\daleth(\alpha), s^{\iota}x) = \alpha \;\cdot\!\!\rightarrow\!\!\cdot \\ & \qquad \cdot\!\!\rightarrow\!\!\cdot\; \aleph_\alpha^{\underline{\aleph_\beta}} = \prod\limits_{x,\daleth(\alpha)} \aleph_{s^{\iota}x}^{\underline{\aleph_\beta}}. \end{cases}
$$

Instead of the exponent $\aleph_\beta$ we could here everywhere write simply an ordinal variable in virtue of $a^{\underline{\nu}} = a^{\underline{\aleph(\nu)}}$.

Concerning the function $\daleth(\alpha)$ we still note the inequality

$$
4.22 \qquad \mathrm{Lim}(\alpha) \rightarrow \aleph_\alpha \subset \aleph_\alpha^{\underline{\daleth(\alpha)}}.
$$

This results from the generalized König theorem 3.16; indeed by this, using also 2.18 and 3.13 we have

$$
\mathrm{Lim}(\alpha) \mathbin{\&} \mathrm{Sqa}(s) \mathbin{\&} \sum_x (\daleth(\alpha), s^{\iota}x) = \alpha \;\cdot\!\!\rightarrow\!\!\cdot
$$

$$
\cdot\!\!\rightarrow\!\!\cdot\; \aleph_\alpha = \sum_x (\daleth(\alpha), \aleph_{s^{\iota}x}) \subseteq \sum_{x,\daleth(\alpha)} \aleph_{s^{\iota}x} \subset \prod_{x,\daleth(\alpha)} \aleph_{s^{\iota}x'} \subseteq \aleph_\alpha^{\underline{\daleth(\alpha)}}.
$$

From 4.22 we especially draw

$$
4.23 \qquad \mathrm{Lim}(\alpha) \mathbin{\&} \daleth(\alpha) \subseteq \beta \rightarrow \aleph_\nu^{\underline{\beta}} \neq \aleph_\alpha.
$$

---

[1]) A newer comprehensive exposition of ordinal and cardinal theory is to be found in [Bachmann 1955].

For, by the premise we have $(\aleph_\nu^{\underline{\beta}})^{\underline{\mathfrak{f}(\alpha)}} = \aleph_\nu^{\underline{\beta}}$, whereas by 4.22 $\aleph_\alpha^{\underline{\mathfrak{f}(\alpha)}} \neq \aleph_\alpha$. A particular consequence of 4.23 is that

$$2^{\underline{\aleph_0}} \; (= \aleph_0^{\underline{\omega}}) \neq \aleph_\omega .$$

As is well known, the question if $2^{\underline{\aleph_0}} = \aleph_{0'}$, as Cantor supposed, is undecided. The assumption that generally $2^{\underline{\aleph_\alpha}} = \aleph_{\alpha'}$, is called the generalized continuum hypothesis. If this relation holds, then obviously many of the foregoing equations and inequalities become considerably simplified. As K. Gödel [1940] has shown, the generalized continuum hypothesis at all events cannot be refuted, provided our system, without the axiom of choice, is consistent.

VII, § 5. Abstract theories

Not only the categorical theories of mathematics, as are number theory, analysis and the theories of ordinals and cardinals, are included in our system, but also the hypothetical theories of abstract algebra and topology can be embodied in it. The method of this embodying was already applied in chap. V for treating order, partial order and lattices. The characteristic means consists in fixing the hypothetical domain and the hypothetical operations and relations by free variables referring to them.

Giving here some instances of this method, we shall content us to show on principle the possibility of embodying the theories in question. For the nearer formal handling it will be required to introduce suitable simplifying conventions.

Let us take as first instance *group theory*. We say that a set a forms a group with respect to the operation f, if f is a functional set with the domain $a \times a$ such that the well known conditions of associativity and invertibility are satisfied.

Formally we define

Df 5.1 $\mathrm{Gr}(f, a) \leftrightarrow \mathrm{Ft}(f) \;\&\; \Delta_1 f = a \times a \;\&\; \Delta_2 f \subseteq a \;\&$
$\&\; (x)(y)(z)(x \in a \;\&\; y \in a \;\&\; z \in a \rightarrow f^{\iota}\langle x, f^{\iota}\langle y, z\rangle\rangle =$
$= f^{\iota}\langle f^{\iota}\langle x, y\rangle, z\rangle) \;\&\; (x)(z)(x \in a \;\&\; z \in a \rightarrow$
$\rightarrow (Ey)(y \in a \rightarrow (f^{\iota}\langle x, y\rangle = z) \;\&\; (Ey)(f^{\iota}\langle y, x\rangle = z)).$

It would be possible to deal also with a function F and a class A constituting a group in a generalized sense, but for formalizing the mathematical reasonings this will not be required, especially if we make use of the axioms A 4 and A 6. If only f is replaced by a function F, then no generalization arises since then the domain of F is $(a \times a)^*$ and thus F is represented.

The formal development of group theory goes by deriving theorems under the premise $\mathrm{Gr}(f, a)$ with the fixed parameters f and a. In the text we shall speak briefly of the group "$\langle f, a\rangle$" and call the elements of a the elements of the group. The order of a group $\langle f, a\rangle$ is simply $\aleph(a)$. The unit element of the group is $\iota_x(f^{\iota}\langle x, x\rangle = x)$, and the inverse element of b is $\iota_x(y)(f^{\iota}\langle f^{\iota}\langle b, x\rangle, y\rangle = y)$. An element c of the group is said to be of infinite or finite order according as the converse domain of the iterator $\mathrm{J}(\{uv \mid v = f^{\iota}\langle u, c\rangle\}, c)$, is infinite or finite. In the last case the cardinal of that converse domain is called the order of c; it then can be expressed by the number term

$$0' + \mu_x(\mathrm{J}(\{uv \mid v = f^{\iota}\langle u, c\rangle\}, c)^{\iota}x = \iota_y(f^{\iota}\langle y, y\rangle = y)).$$

The definition of a subgroup of $\langle f, a\rangle$ is given by

Df 5.2 $\quad \mathrm{Sgr}(b, f, a) \leftrightarrow \mathrm{Gr}(f, a) \;\&\; b \subseteq a \;\&\; \mathrm{Gr}(f \cap ((b \times b) \times a), b)$

«b is a subgroup of a with respect to the operation f».

An example of an higher concept related to group theory is that of the group of automorphisms of a set c with respect to a certain—for instance ternary—relation H, which is a subclass of $(V \times V) \times V$. Then the class of automorphisms of c with respect to H is

$$\{x \mid \mathrm{Crs}(x) \;\&\; \Delta_1 x = c \;\&\; \Delta_2 x = c \;\&\; (u)(v)(w)(u \in c \;\&\; v \in c \;\&\; w \in c \cdot\!\!\rightarrow\!\!\cdot$$
$$\cdot\!\!\rightarrow\!\!\cdot \langle\langle u, v\rangle, w\rangle \in H \leftrightarrow \langle\langle x^{\iota}u, x^{\iota}v\rangle, x^{\iota}w\rangle \in H)\}.$$

This class is represented by a set t, since it is a subclass of $\pi(c \times c)$; and t is a group with respect to the function

$$\{uv \mid (Ex)(Ey)(x \in t \;\&\; y \in t \;\&\; u = \langle x, y\rangle \;\&\; v = x|y)\},$$

which is represented, since its domain is $t \times t$.

From these brief indications it will appear that group theory can be developed in the frame of our system. There exists the class of all groups, or in other words the class of all models for the axioms of group theory, namely $\{xy \,\|\, \mathrm{Gr}(x, y)\}$.

Especially near to set theory is the *topology* of *spaces*. Let us briefly consider the method of starting in topology with a concept of closure. Generally a functional set c is said to be a closure function for a set a, if it assigns to every subset of a again a subset in such a way that the following closure conditions are fulfilled:

Df 5.3 $$\mathrm{Cl}(c, a) \leftrightarrow \mathrm{Ft}(c) \;\&\; \Delta_1 c = \pi(a) \;\&\; \Delta_2 c \subseteq \pi(a) \;\&$$ $$\& \;(x)(y)(x \subseteq a \;\&\; y \subseteq a \;\cdot\!\!\rightarrow\!\cdot$$ $$\cdot\!\!\rightarrow\!\cdot\; x \subseteq c'x \;\&\; c'c'x \subseteq c'x \;\&\; (x \subseteq y \rightarrow c'x \subseteq c'y)).$$

Now a set s is said to be a space with respect to a closure function c if

$$\mathrm{Cl}(c, s) \;\&\; (x)(y)(x \subseteq s \;\&\; y \subseteq s \rightarrow c'(x \cup y) = c'x \cup c'y) \;\&$$ $$\& \;(x)(x \in s \rightarrow c'[x] = [x]).$$

Then the point sets to be considered are the subsets of s, and hence the language of the theory is directly set theoretic. The derivations are expressible quite in the familiar way. Of course on a certain stage of the theory, namely when upon suitable conditions a distance function is to be defined, we have to adjoin the theory of real numbers.

Somewhat more complicated is the embodying of such theories where formal operating and abstract reasonings are combined, as it occurs in the *algebraic theory* of *rings*. For this case a method of that embodying procedure shall be briefly indicated.

We begin with the concept of a ring. Restricting our consideration to commutative rings with a multiplicative unit we introduce by explicit definition the predicate with five free arguments $\mathrm{Rg}(f, g, t, o, e)$, to read "$t$ is with respect to f as ring sum and g as ring product a commutative ring with null element o and unit element e ($\neq o$)". It will not be necessary to write down here this definition which formulates the familiar properties of such a ring. In the text we shall speak briefly of "the ring $\langle\langle f, g\rangle, t\rangle$". (Note that o and e are determined by f, g, t.) We also take the liberty to speak of an element of this ring in the sense of an element of t.

There are some of the elementary notions connected with the concept of a ring which we shall have to use. The k-fold of a ring element a, k being a natural number, is defined by

$$\mathsf{J}(\{uv \mid v = f`\langle u, a\rangle\}, o)`k$$

and likewise the kth potency of a ring element a by

$$\mathsf{J}(\{uv \mid v = g`\langle u, a\rangle\}, e)`k.$$

The *ring sum* Σs of the members of a finite sequence s of ring elements and likewise the *ring product* Πs of these members is definable by primitive recursion. (Cf. III, 4.7).

Now we go on to define a polynomial over our ring $\langle\langle f, g\rangle, t\rangle$ in n variables $x_0, x_1 \ldots x_{n-1}$. Intuitively it consists of Potenzprodukte of the n variables with coefficients out of $t^{-}[o]$, joined by "+". Each Potenzprodukt is determined by the powers assigned to the different variables. Thus we can formally represent it by a sequence of natural numbers with the domain n (n-sequence of natural numbers), which we call briefly a *n-complex*. Correspondingly for a Potenzprodukt with coefficient we take an ordered pair whose first member is a n-complex and the second member an element of $t^{-}[o]$. We call such a pair a monomial and the second member the coefficient of the monomial. Now a *polynomial* (n-polynomial over our ring) is defined as a finite functional set whose elements are monomials. By this definition the empty set is a particular n-polynomial. The coefficients of the monomials in a polynomial are also called the coefficients of this polynomial. Equality of polynomials is set-theoretic identity.

We state at once some familiar notions, related to n-complexes and polynomials. By the degree of a n-complex we understand the number-theoretic sum of its members, and by the degree of a polynomial the highest degree of a complex of its domain. (The degree of the empty polynomial is taken as zero.) A polynomial is called homogenious if the n-complexes of its domain have all the same degree. Two n-complexes p, q are said to be associated if there is a one-to-one correspondence c of n into itself (permutation) such that $c \mid p = q$. A polynomial is called symmetrical if any two of its monomials with associated n-complexes have the same coefficient.

We now come to define algebraic sum and product of polynomials.

The algebraic sum of polynomials p, q is obtained as follows. We first form the set of monomials whose domain is the set sum of the domains of p and q and whose each complex c occurs with that coefficient which is either—in case that c occurs only in one of the polynomials p, q—$p^{\prime}c$ or $q^{\prime}c$, and in case that c occurs in both, $f^{\prime}\langle p^{\prime}c, q^{\prime}c\rangle$. From this set we get the algebraic sum of p and q by cancelling the monomials with the coefficient o. The associativity and commutativity of this sum can be inferred from the corresponding properties of the ring sum. For the so defined sum operation on polynomials the null element is the empty polynomial, and the opposite of a polynomial p is that polynomial which assigns to every complex c of the domain of p that element of the ring which is opposite to $p^{\prime}c$. Still the sum of the members of a finite sequence of polynomials can be defined in the usual manner by primitive recursion.

In order to define the algebraic product of polynomials we first define the product of two n-complexes c and d as that n-complex which assigns to any natural number $k \in n$ the natural number $c^{\prime}k + d^{\prime}k$. The product of two monomials $\langle c, a\rangle$, $\langle d, b\rangle$ is the monomial whose first member is the product of the complexes c and d and whose second member is $g^{\prime}\langle a, b\rangle$.[1] The product of a monomial r with a polynomial q is that polynomial whose elements are the products of r with the elements of q. The algebraic product of two polynomials p, q now is got as follows. Let h be a numeration of p (which is a finite sequence). We pass from h to that sequence s, which assigns to each natural number l out of the domain of h the product of $h^{\prime}l$ with q. The sequence s is then a sequence of poly-

[1]) However in the case that $g^{\prime}\langle a, b\rangle$ is 0, the product of the monomials "vanishes"; it then has simply to be dropped in the following product formation.

nomials and the algebraic sum of its members can be proved to be independent from the special numeration of p. This algebraic sum then is the algebraic product of p and q.

The proofs of the commutativity and associativity of this product as also of the distributive law have to be made by means of numeral inductions with respect to the multitudes of the occurring polynomials applying the corresponding properties of the ring operations and the commutativity and associativity of products of monomials. The unit element for the algebraic product operation is the polynomial $[\langle n \times [o], e\rangle]$. We note also that an algebraic product of which one factor is the empty polynomial yields the empty polynomial.

On the whole we thus get that the polynomials over our ring—the "coefficient ring"—constitute with respect to the defined algebraic sum and product operations, again a commutative ring with unit element.

In this ring to every element a of $t^{-}[o]$ corresponds a constant polynomial $[\langle n \times [o], a\rangle]$; these polynomials together with the empty polynomial, taken as corresponding to o, form a subring which is isomorphic to our ring $\langle\langle f, g\rangle, t\rangle$.

For the following we note still that we can define the product of a polynomial p with an element a of t, namely as its product with the constant polynomial corresponding to a; this product is to be obtained from p by replacing every coefficient c of it by $g^{t}\langle a, c\rangle$; only in case that a is o the product is the empty set.

An essential procedure in the theory of polynomials is to substitute for the variables elements of a ring r. For the possibility of this process it is required that the product of an element of r with an element of the coefficient ring can be defined as an element of r. The substitution process for a polynomial p is then as follows. First in every element $\langle l, k\rangle$ of a n-complex c of p we replace the natural number k by the kth potency of that element of r which is to be substituted for the variable with the number l. By this we get from the complex c a sequence s of elements of r; and its ring product Πs is the value to be substituted for c. Further for every monomial $\langle c, a\rangle$ of p the value to be substituted is the product of the value substituted for c with the element a of the coefficient ring. Now we take a numeration of p; replacing here every monomial in the converse domain by the substituted value and forming the ring sum Σs of the resulting sequence s we obtain the value to be substituted for p.

As an instance for the application of this substitution process we formulate in our frame the theorem on symmetrical n-polynomials over $\langle\langle f,g\rangle,t\rangle$ which says that every such polynomial is expressible as a polynomial over $\langle\langle f,g\rangle,t\rangle$ in the elementary symmetrical n-polynomials.

We define the elementary symmetrical polynomial with number l ($l \in n$) to be the set of those pairs $\langle c, e\rangle$ wherein c is an n-complex associated to the set representing

$$\{xy \mid x \in l' \,\&\, y = 0' \cdot \mathbf{v} \cdot x \in n^{-}l' \,\&\, y = 0\}.$$

(Note that we begin the numbering of elementary symmetrical polynomials, like that of the variables, with null.)

Now in order to state the theorem we need only still to express what it means that a n-polynomial p is expressible in n special numbered polynomials. This can be formulated by saying that the polynomial p results as the value obtained from substituting the numbered polynomials for the equally numbered variables in a certain n-polynomial. The substitution can be performed, since the elementary symmetrical polynomials are elements of the ring of n-polynomials over the ring $\langle\langle f,g\rangle,t\rangle$ and the product of an n-polynomial with an element of this ring has been defined as an n-polynomial. –

These instances might be suffice to make appear the possibility of englobing formal algebra in our set-theoretic frame. We have refrained here from a more explicit formalization because this would have required to introduce a lot of notations what seemed not to be worthwhile, as we are not going farther into the subject. On the other hand the definitions here given are formulated in such a way that the passage to formulas of our system is at hand.

CHAPTER VIII

FURTHER STRENGTHENING OF THE AXIOM SYSTEM

VIII, § 1. A STRENGTHENING OF THE AXIOM OF CHOICE

In this chapter we are considering certain extensions of our axiomatic system which, as the preceding chapters show, are not required for establishing classical mathematics, but have the effect of giving the system a stronger closure. This will be attained by strengthening the axiom of choice and adopting a form of the "Fundierungsaxiom", as called by Zermelo [1930].

First as to axiom of choice, a motive for modifying the form A 5 hitherto used can be found in the circumstance that this form of the axiom is in a striking way deviating from the explicit form of all our other axioms A 1–A 4, A 6. Indeed with each of these axioms a new function symbol $\mathfrak{f}$ or individual symbol $\mathfrak{a}$ is introduced, by which a "symbolische Auflösung" of an existential statement $(x_1)\ldots(x_n)(Ey)\mathfrak{A}(x_1\ldots x_n, y)$ is afforded in the form

$$(x_1)\ldots(x_n)\mathfrak{A}(x_1\ldots x_n, \mathfrak{f}(x_1\ldots x_n))$$

or (if n is zero) $\mathfrak{A}(\mathfrak{a})$ [1]).

The question thus arises if we cannot apply this method to one of the different forms of the axiom of choice stated in VI, § 2. This indeed is possible. We perform the symbolische Auflösung with starting from the formula VI, 2.5. Here by introducing the function symbol $\sigma(a, b)$, and cancelling the conjunction members $\mathrm{Ft}(y)$ and $\Delta_1 y = m$, which now become redundant, we get first

[1] $$(x)(x \in m \rightarrow x \neq 0) \cdot\!\rightarrow\!\cdot\, c \in m \rightarrow \sigma(m, c) \in c.$$

Now we observe that [1] can be replaced by a simpler formula; namely by substituting $[c]$ for m and dropping the redundant antecedent $c \in [c]$ the formula [1] gives

[2] $$(x)(x \in [c] \rightarrow x \neq 0) \rightarrow \sigma([c], c) \in c.$$

---

1) Concerning the concept of symbolische Auflösung cf. [Hilbert and Bernays 1934/39], vol. II, § 1, p. 6.

Combining [2] with the obvious formula

$$c \neq 0 \rightarrow (x)(x \in [c] \rightarrow x \neq 0)$$

we obtain

[3] $$c \neq 0 \rightarrow \sigma([c], c) \in c.$$

Now defining $\sigma(c) = \sigma([c], c)$, we come to the pregnant formula

A_σ $$c \neq 0 \rightarrow \sigma(c) \in c,$$

which is an explicit form of the axiom of choice. Indeed from A_σ we come back to VI, 2.5 by means of the formula

[4] $$(x)(x \in m \rightarrow x \neq 0) \;\&\; b^* \equiv \{xz \mid x \in m \;\&\; z = \sigma(x)\} \cdot\rightarrow\cdot$$
$$\cdot\rightarrow\cdot \mathrm{Ft}(b) \;\&\; \Delta_1 b = m \;\&\; (x)(x \in m \rightarrow b^{\iota}x \in x),$$

using II, 3.10.

An inverse passage is not possible. In fact, E. Specker proved that A_σ cannot be inferred (using a definition of $\sigma(a)$) from A 5 and our other axioms, provided our system is consistent.

Thus A_σ is a strengthened form of the axiom of choice. That A_σ is not directly included in A 5 appears from the interpretation. Indeed A_σ expresses the existence of a function which assigns to every non empty set one of its elements. Such a function is a subclass of the class $\{xy \mid y \in x\}$ with the same domain. By A 5 the existence of such a function is stated only for sets of pairs.

With regard to our application of the symbolische Auflösung two things may be observed. The one is that this process itself amounts to applying a form of a choice principle. Therefore in using this process with respect to the statement of a choice axiom two applications of a choice principle so to speak are superposed. The second remark is that the strength of our application of that process, in introducing $\sigma(a, b)$, is due to the circumstance that the variable m in VI, 2.5 ranges over the class of all sets, and not only over a set.

The elegant explicit form A_σ of the axiom of choice was first given by Th. Skolem [1929]. In our system it removes the above mentioned deviation in structure of the axiom of choice from the other axioms. On the other hand we must be conscious of a new kind of deviation from the other axioms which here arises. Indeed all our former axioms A 1–A 4, A 6, include an extensional definition of the symbol introduced by it. This however is obviously not the case for the symbol σ. Nevertheless $\sigma(a)$ syntactically figures as a function, for which we also can derive by means of the equality axioms the formula

1.1 $$a = b \rightarrow \sigma(a) = \sigma(b).$$

The adoption of A_σ instead of A 5 allows in particular to make dispensable in several proofs the application of A 4.

Thus in the proof of the numeration theorem VI, 3.1 we have only to take for G the function $\{xy \mid y = \sigma(c^- x)\}$. – Further in the proof of the Hausdorff principle VI, 4.1 the function g there occurring can be replaced by

$$\{uv \mid u \subset \Delta_1 c \;\&\; (Ex)(\mathfrak{C}(x, u) \;\&\; v \;=\; \sigma(\{x \mid \mathfrak{C}(x, u)\} \cap \Delta_1 c)\}.$$

In fact it is not used in the proof that g is a set.

So we find that in the passage from the axiom of choice to the Hausdorff principle and thus also to Zorn's lemma the application of A 4 is dispensable, if we use the form A_σ of the choice axiom instead of A 5. However for the inverse passage from Zorn's lemma to A 5 it seems that the application of A 4 cannot be avoided [1]).

There is still a stronger form of the choice principle, suggested by the Hilbert "ε-formula"

$$C(a) \rightarrow C(\varepsilon_x C(x))$$

which is translated in our frame by taking a class variable instead of the predicate variable and $\sigma(C)$ instead of $\varepsilon_x C(x)$. In this way we come to the axiom

A'_σ $$a \in C \rightarrow \sigma(C) \in C.$$

The parallelism of A'_σ with A_σ is obvious. Besides A_σ is derivable from A'_σ upon defining $\sigma(a) = \sigma(a^*)$. However the formula which would correspond to 1.1 is not provable; namely the schema I, 3.1 is no more derivable after $\sigma(A)$ being introduced, since the axiom

[1]) Cf. the corresponding remark, relative to the simple type theory, by R. Büchi [1953].

A'_σ gives no means for eliminating this symbol. Therefore in order to complete the parallelism between $\sigma(A)$ and $\sigma(a)$ we have to add the axiom

A''_σ $$A \equiv B \rightarrow \sigma(A) = \sigma(B).$$

In a like way as A'_σ arises from A_σ by changing a set variable into a class variable, there is a strengthening of A 5 for the case of the set a being replaced by a class A. This strengthened statement can be derived from A'_σ; namely using this axiom we have

1.2 $$\mathrm{Ps}(A) \,\&\, F \equiv \{uv \mid u \in \Delta_1 A \,\&\, v = \sigma(\{x \mid \langle u, x\rangle \in A\})\} \rightarrow \\ \rightarrow F \subseteq A \,\&\, \Delta_1 F \equiv \Delta_1 A \,\&\, \mathrm{Ft}(F).$$

In comparison with A 5 there is here the difference that instead of the pure existential assertion we have an explicit indication of the class stated to exist. The pure existential statement was taken in [Bernays 1941] as axiom of choice. In our present frame it cannot be directly formalized, since we do not employ bound class variables.

The axiom A'_σ affords a general method for formalizing the symbolische Auflösung. Let us show this for the case of the passage from $(x)(Ey)\mathfrak{A}(x, y)$ to $(x)\mathfrak{A}(x, \mathfrak{f}(x))$ – with a new function symbol $\mathfrak{f}$. Indeed here A'_σ together with the Church schema and the predicate calculus gives

$$\begin{aligned} \mathfrak{A}(a, b) &\rightarrow b \in \{z \mid \mathfrak{A}(a, z)\} \\ &\rightarrow \sigma(\{z \mid \mathfrak{A}(a, z)\}) \in \{z \mid \mathfrak{A}(a, z)\} \\ &\rightarrow \mathfrak{A}(a, \sigma(\{z \mid \mathfrak{A}(a, z)\})) \end{aligned}$$

and hence

$$(x)(Ey)\mathfrak{A}(x, y) \rightarrow (x)\mathfrak{A}(x, \sigma(\{z \mid \mathfrak{A}(x, z)\})),$$

so that upon our premise we get $(x)\mathfrak{A}(x, \mathfrak{f}(x))$ by defining

$$\mathfrak{f}(c) = \sigma(\{z \mid \mathfrak{A}(c, z)\}).$$

From A'_σ together with A''_σ we still can derive a further choice principle, which is a generalization of the multiplicative axiom VI, 2.6 [1]);

[1]) This form of the choice principle was stated by E. Specker in his Habilitationsschrift [1957].

1.3 $\mathrm{Ps}(A)$ & $(x)(y)(z)(\langle x, z\rangle \in A$ & $\langle y, z\rangle \in A \rightarrow \langle x, y\rangle \in A$ & & $\langle z, z\rangle \in A)$ & $C \equiv \{y \mid (Ex)(y = \sigma(\{z \mid \langle x, z\rangle \in A\}))\} \cdot\!\!\rightarrow\!\!\cdot$ $\cdot\!\!\rightarrow\!\!\cdot\ (x)(x \in \Delta_1 A \rightarrow (Ey)(y \in C$ & $\langle x, y\rangle \in A$ & $(z)(z \in C$ & & $\langle x, z\rangle \in A \rightarrow y = z)))$

«If A is a class of pairs such that by $\langle a, b\rangle \in A$ the domain $\Delta_1 A$ is divided in mutually exclusive classes, each expressible in the form $\{z \mid \langle a, z\rangle \in A\}$, then there exists a class C which has exactly one element in common with each of these classes».

The proof goes as follows. Denoting by $\mathfrak{s}(a)$ the term

$$\sigma(\{z \mid \langle a, z\rangle \in A\})$$

we have in virtue of A'_σ and the Church schema

[1] $$a \in \Delta_1 A \rightarrow \langle a, \mathfrak{s}(a)\rangle \in A.$$

It will be sufficient to derive from our premises the formula

$$a \in \Delta_1 A \rightarrow \mathfrak{s}(a) \in C \;\&\; \langle a, \mathfrak{s}(a)\rangle \in A \;\&\; (c \in C \;\&\; \langle a, c\rangle \in A \rightarrow \mathfrak{s}(a) = c).$$

From the third premise we get, using again A'_σ

[2] $$b \in C \leftrightarrow (Ex)(x \in \Delta_1 A \;\&\; b = \mathfrak{s}(x))$$

from which we also infer

[3] $$a \in \Delta_1 A \rightarrow \mathfrak{s}(a) \in C.$$

By [1], [2] and [3] it only remains to derive

$$d \in \Delta_1 A \;\&\; \langle a, \mathfrak{s}(d)\rangle \in A \rightarrow \mathfrak{s}(a) = \mathfrak{s}(d),$$

what by [1] comes out to prove

$$\langle d, \mathfrak{s}(d)\rangle \in A \;\&\; \langle a, \mathfrak{s}(d)\rangle \in A \rightarrow \mathfrak{s}(a) = \mathfrak{s}(d).$$

For this we employ our first premise, by which we get successively

$$\begin{aligned} &\langle d, \mathfrak{s}(d)\rangle \in A \;\&\; \langle a, \mathfrak{s}(d)\rangle \in A \rightarrow \langle a, d\rangle \in A \\ &\qquad \rightarrow \{z \mid \langle a, z\rangle \in A\} \equiv \{z \mid \langle d, z\rangle \in A\} \\ &\qquad \rightarrow \sigma(\{z \mid \langle a, z\rangle \in A\}) = \sigma(\{z \mid \langle d, z\rangle \in A\}) \quad \text{by } A''_\sigma \\ &\qquad \rightarrow \mathfrak{s}(a) = \mathfrak{s}(d). \end{aligned}$$

With the help of A'_σ we now still can prove an assertion stated in the beginning of V, § 3, whose proof then was delated, namely that for every non empty ordered class A with no foremost element there is a subset with the same property:

$$1.4 \quad \mathrm{Or}(C) \;\&\; A \subseteq \Delta_1 C \;\&\; a \in A \;\&\; (x)(x \in A \rightarrow \rightarrow (Ey)(y \in A \;\&\; y \neq x \;\&\; \langle y, x\rangle \in C)) \cdot\!\rightarrow\!\cdot (Ez)(z^* \subseteq A \;\&\; z \neq 0 \;\&\; (x)(x \in z \rightarrow (Ey)(y \in z \;\& \;\&\; y \neq z \;\&\; \langle y, x\rangle \in C))).$$

For the proof we denote by $\mathfrak{K}$ the iterator

$$\mathrm{J}(\{uv \mid u \in A \;\&\; v = \sigma(\{y \mid y \in A \;\&\; y \neq u \;\&\; \langle y, u\rangle \in C\})\}, a).$$

In virtue of our premise we have by III, 4.1 and A 6

$$\mathrm{Ft}(\mathfrak{K}) \;\&\; \Delta_1\mathfrak{K} \equiv \omega^* \;\&\; \Delta_2\mathfrak{K} \subseteq A \;\&\; (n \in \omega \rightarrow \rightarrow \mathfrak{K}`n' \neq \mathfrak{K}`n \;\&\; \langle \mathfrak{K}`n', \mathfrak{K}`n\rangle \in C).$$

Now, by II, 3.8

$$(Ez)(z^* \equiv \Delta_2\mathfrak{K})$$

and thus the succedent of 1.4 results.

In the following we shall show that A'_σ and A''_σ become derivable from A_σ, under a suitable definition of $\sigma(A)$, if we add the Fundierungsaxiom.

VIII, § 2. The fundierungsaxiom

The idea of adding the Fundierungsaxiom to the axioms of set theory goes back to Mirimanoff [1917–1920], who took it as a means for excluding some sets of an anomalous kind. In fact by our axioms the concept of a set is precized only in the sense that definite processes of set formation are axiomatically introduced. But by this way the possibility of some anomalies is not excluded, which lie beyond the ordinary mathematical set formations; on the other hand this possibility is a hindering for some proofs. An instance of such an anomaly is a set which is its own element, or else a set $[a, b]$ such that each of its elements is an element of the other. In the theory of ordinals we have excluded such anomalies by requiring ordinals to satisfy the property Fund (III, Df 1.3).

Thus we are lead to postulating generally Fund(a) for every set a, what can be stated by the formula

$$b \subseteq a \;\&\; b \neq 0 \;\rightarrow\; (Ey)(y \in b \;\&\; y \cap b = 0).$$

An equivalent statement obviously is

A 7 $\qquad c \neq 0 \;\rightarrow\; (Ey)(y \in c \;\&\; (x)(x \notin y \vee x \notin c))$

«For every non empty set c there is an element which has no element in common with c».

From the way we have come to A 7 it is evident that, if A 7 is present, the definition of Od (III, Df 1.4) can be simplified to

2.1 $\qquad \mathrm{Od}(d) \leftrightarrow \mathrm{Trans}(d) \;\&\; \mathrm{Alt}(d).$

Applying A 7 to the sets $[a]$, $[a, b]$, $[a, b, c]$ we get

2.2 $\qquad a \notin a, \quad a \in b \rightarrow b \notin a, \quad a \in b \;\&\; b \in c \rightarrow c \notin a.$

Generally we can prove with A 7

2.3 $\qquad \mathrm{Sq}(s) \;\&\; \Delta_1 s = k' \;\&\; \mathrm{Nu}(k) \;\&\; (x)(x \in k \rightarrow$

$$\rightarrow s`x \in s`x') \;\rightarrow\; s`k \notin s`0.$$

This goes by deriving, upon our premises, using the number-theoretic formula

$$\mathrm{Nu}(k) \cdot\!\rightarrow\!\cdot (x)(x \in k' \;\&\; x \neq 0 \;\rightarrow\; (Ey)(y \in k \;\&\; x = y')),$$

the formula

$$(x)(x \in k' \;\&\; x \neq 0 \;\rightarrow\; s`x \cap \Delta_2 s \neq 0),$$

which gives together with A 7 applied to $\Delta_2 s$

$$s`0 \cap \Delta_2 s = 0$$

and hence

$$s`k \notin s`0.$$

The consequence 2.3 of A 7 excludes the existence of finite sets whose elements are in a cyclic element relation. But A 7 also excludes the existence of ω-sequences whose every member has the following member as element. In fact we have (cf. VI, Df 5.4)

2.4 $$\overline{(Ey)}\ (\mathrm{Dsq}(y)\ \&\ (x)(x \in \omega \rightarrow y^{\iota}x' \in y^{\iota}x)),$$

since

[1] $$\mathrm{Dsq}(s)\ \&\ (x)(x \in \omega \rightarrow s^{\iota}x' \in s^{\iota}x) \rightarrow (z)(z \in \Delta_2 s \rightarrow z \cap \Delta_2 s \neq 0)$$

and the succedent of [1] contradicts A 7.

With the aid of this theorem we can prove, using A_σ, the stronger statement of the Fundierungsaxiom which was adopted by von Neumann [1929], Zermelo [1930] and in [Bernays 1941][1]):

2.5 $$a \in C \rightarrow (Ey)(y \in C\ \&\ y \cap C = 0).$$

The proof goes indirectly by deriving a contradiction from the premise (with fixed a and C)

[1] $$a \in C\ \&\ (y)(y \in C \rightarrow y \cap C \neq 0).$$

For this we take the iterator

$$\mathrm{J}(\{xy \mid y = \sigma(x \cap C)\}, a)$$

which by A 6 and II, 3.10 is represented by an ω-sequence s. Now using our premise [1] we first infer $a \cap C \neq 0$, i.e. $s^{\iota}0 \cap C \neq 0$, and from the characterizing property of the iterator it follows by numeral induction on n (using again [1])

$$\mathrm{Nu}(n) \rightarrow s^{\iota}n' \in s^{\iota}n \cap C.$$

But this contradicts 2.4.

The application of A_σ in this proof can be avoided by a method which uses A 4. This reasoning, due to Gödel [1940], will be given in the next section.

We still observe that in the case we have the axiom A_σ at our disposal, the Fundierungsaxiom can be taken in the simple form

F_σ $$\sigma(a) \cap a = 0.$$

---

[1]) As was carried out in [Bernays, 1941] with indication of literature, ordinal theory, so far as developed in III, § 1, can be obtained with taking as axioms A 1, A 2 and the formula 2.5. The method here used has afterwards been found to be mainly contained already in Mirimanoff [1917].

In fact from this formula together with A_σ, A 7 directly follows.

The corresponding formula, related to $\sigma(A)$,

$$F'_\sigma \qquad \sigma(A) \cap A = 0$$

can be reduced, as we shall see, to F_σ in an analogous way as A'_σ is derivable from A_σ.

VIII, § 3. A ONE-TO-ONE CORRESPONDENCE BETWEEN THE CLASS OF ORDINALS AND THE CLASS OF ALL SETS

The axioms A_σ and F_σ can be applied to set up a one-to-one correspondence between the class of ordinals and the class of all sets. For this purpose we start from a construction, due to von Neumann, which is still independent of A_σ and F_σ.

We define

Df 3.1 $$\Psi \equiv \mathrm{I}(\{xy \mid y = \pi(x)\}, 0).$$

Then we have

3.1 $$\mathrm{Ft}(\Psi),\ \Delta_1\Psi \equiv \{x \mid \mathrm{Od}(x)\}$$

3.2 $$\mathrm{Trans}(\Psi`\alpha).$$ [1]

The proof of 3.2 goes by transfinite induction. We have first

[1] $$\mathrm{Trans}(0),$$

further

[2] $$\mathrm{Trans}(a) \rightarrow \mathrm{Trans}(\pi(a)),$$

which follows by

$$(x)(x \in a \rightarrow x \subseteq a) \rightarrow (x)(z)(z \subseteq a \mathbin{\&} x \in z \rightarrow x \subseteq a)$$
$$\rightarrow (z)(z \in \pi(a) \rightarrow z \subseteq \pi(a)),$$

and obviously also

[3] $$(\mathfrak{x})(\mathfrak{x} \in a \rightarrow \mathrm{Trans}(\mathfrak{t}(\mathfrak{x}))\cdot\rightarrow\cdot\mathrm{Trans}(\sum_{\mathfrak{x}}(a, \mathfrak{t}(\mathfrak{x}))).$$

By [1], [2], [3] and Df 3.1 transfinite induction yields 3.2.

---

[1] As in chapt. VII, §§ 2–4 we use in this section specialized variables for ordinals.

Moreover we get

3.3 $$\Psi`0=0,\ \Psi`\alpha \in \Psi`\alpha',$$

and by 3.2

3.4 $$\Psi`\alpha \subseteq \Psi`\alpha'.$$

Further, since $\mathrm{Fin}(a) \rightarrow \mathrm{Fin}(\pi(a))$, we have

3.5 $$\mathrm{Nu}(\alpha) \rightarrow \mathrm{Fin}(\Psi`\alpha).$$

By VI, 1.10 we get

3.6 $$\aleph(\Psi`\alpha) \in \aleph(\Psi`\alpha'),$$

which together with 3.4 gi

3.7 $$\Psi`\alpha \subset \Psi`\alpha'.$$

Further from

$$\mathrm{Lim}(\lambda)\ \&\ \alpha \in \lambda \rightarrow \Psi`\alpha' \subseteq \Psi`\lambda$$

together with 3.3, 3.6, 3.7 we get

$$\mathrm{Lim}(\lambda)\ \&\ \alpha \in \lambda \rightarrow \Psi`\alpha \in \Psi`\lambda\ \&\ \Psi`\alpha \subset \Psi`\lambda\ \&\ \aleph(\Psi`\alpha) \in \aleph(\Psi`\lambda).$$

Hence by transfinite induction results

3.8 $$\alpha \in \beta \rightarrow \Psi`\alpha \in \Psi`\beta\ \&\ \Psi`\alpha \subset \Psi`\beta\ \&\ \aleph(\Psi`\alpha) \in \aleph(\Psi`\beta).$$

We now define

Df 3.2 $$\Pi \equiv \bigcup \Delta_2 \Psi,$$

by which we have

3.9 $$a \in \Pi \leftrightarrow (E\xi)(a \in \Psi`\xi).$$

We shall call an element of Π briefly a Π-set. The class Π, as a union of transitive sets, is a transitive class, i.e. every element of a Π-set is itself a Π-set:

3.10 $$a \in b\ \&\ b \in \Pi \rightarrow a \in \Pi.$$

The Π-sets are characterized as those which occur as elements in some $\Psi`\alpha$; and the least $\alpha$ for which a $\Pi$-set $a$ is in $\Psi`\alpha$ is called the *degree* of a, formally

Df 3.3 $$\mathrm{dg}(a) = \mu_x(a \in \Psi`x).$$

From this definition together with Df 3.1 the following properties of dg(a) result:

3.11 $a \notin \Pi \rightarrow \mathrm{dg}(a)=0, \quad a \in \Pi \rightarrow \mathrm{Suc}(\mathrm{dg}(a))$

3.12 $a \in \Pi \rightarrow a \in \Psi^{\iota}\mathrm{dg}(a) \;\&\; (\xi)(a \in \Psi^{\iota}\xi \leftrightarrow \mathrm{dg}(a) \subseteq \xi)$

3.13 $a \in b \;\&\; b \in \Pi \rightarrow \mathrm{dg}(a) \in \mathrm{dg}(b)$;

the last is proved in the following way:

$$\begin{aligned} a \in b \;\&\; \mathrm{dg}(b)=\alpha' &\rightarrow b \in \Psi^{\iota}\alpha' \\ &\rightarrow b \in \pi(\Psi^{\iota}\alpha) \\ &\rightarrow b \subseteq \Psi^{\iota}\alpha \\ &\rightarrow \mathrm{dg}(a) \subseteq \alpha. \end{aligned}$$

By means of the concept dg(a) we still prove that every set of Π-sets is itself a Π-set:

3.14 $$c^* \subseteq \Pi \rightarrow c \in \Pi.$$

For this we consider the class

$$\{uv \,\|\, u \in c \;\&\; v=\mathrm{dg}(u)\}.$$

This is a function with the domain c, thus the converse domain is represented by a set d. Since d is a set of ordinals, $\sum d$ is an ordinal ν. Moreover we have

$$(x)(x \in c \rightarrow \mathrm{dg}(x) \subseteq \nu),$$

hence by 3.12

$$c \subseteq \Psi^{\iota}\nu$$
$$c \in \Psi^{\iota}\nu'.$$

Now the existence of a one-to-one correspondence between $\{x \,\|\, \mathrm{Od}(x)\}$ and Π results in the following way, using A_σ.

First we have in virtue of 3.5 – 3.7 and VII, 4.4

3.15 $$\aleph(\Psi^{\iota}\alpha')\text{-}\aleph(\Psi^{\iota}\alpha) \sim \Psi^{\iota}\alpha'\text{-}\Psi^{\iota}\alpha.$$

Thus for every α the class C of one-to-one correspondences between $\aleph(\Psi^{\iota}\alpha')\text{-}\aleph(\Psi^{\iota}\alpha)$ and $\Psi^{\iota}\alpha'\text{-}\Psi^{\iota}\alpha$ is not empty. On the other hand each of these one-to-one correspondences is $\subseteq \aleph(\Psi^{\iota}\alpha') \times \Psi^{\iota}\alpha'$, therefore $C \subseteq \pi(\aleph(\Psi`\alpha') \times \Psi`\alpha)^*$ and hence C is represented by a set, which is expressed by the ι-term:

$$\iota_x(x^* \equiv \{z \,|\, \mathrm{Crs}(z) \;\&\; \Delta_1 z = \aleph(\Psi`\alpha') \dot{-} \aleph(\Psi`\alpha) \;\&\; \Delta_2 z = \Psi`\alpha' - \Psi`\alpha\}.$$

Denoting it by $\mathfrak{S}(\alpha)$ we have by 3.15

$$\mathfrak{S}(\alpha) \neq 0$$

and hence by A_σ

$$\sigma(\mathfrak{S}(\alpha)) \in \mathfrak{S}(\alpha).$$

Now we define

Df 3.4 $$\Theta \equiv \{u \,|\, (E\xi)(u \in \sigma(\mathfrak{S}(\xi)))\}.$$

Θ is evidently a pairclass; further it is a one-to-one correspondence. This results from the formulas

[1] $\alpha \in \beta \rightarrow \Psi`\alpha' - \Psi`\alpha \cap \Psi`\beta' - \Psi`\beta = 0$

[2] $\alpha \in \beta \rightarrow \aleph(\Psi`\alpha') \dot{-} \aleph(\Psi`\alpha) \cap \aleph(\Psi`\beta') \dot{-} \aleph(\Psi`\beta) = 0,$

which follow by $\alpha \in \beta \rightarrow \alpha' \subseteq \beta$ and 3.8.

The converse domain of Θ is Π. Namely first for every α, $(\Psi`\alpha')^* \subseteq \Pi$, and on the other hand every element $c$ of $\Pi$ has a degree which is a successor $\alpha'$ such that $c \in \Psi`\alpha' - \Psi`\alpha$. The domain of $\Theta$ is $\{x \,|\, \mathrm{Od}(x)\}$. For showing this it is sufficient, in virtue of $\alpha \subseteq \aleph(\Psi`\alpha)$, to prove

[3] $\nu \in \aleph(\Psi`\alpha) \rightarrow (E\xi)(\xi \in \alpha \;\&\; \nu \in \aleph(\Psi`\xi') \dot{-} \aleph(\Psi`\xi)).$

This goes by transfinite induction with respect to α. Let [3] be satisfied with α replaced by any $\varkappa$ lower than α. In the cases $\alpha = 0$ and $\mathrm{Suc}(\alpha)$ [3] is simply proved. In the case $\mathrm{Lim}(\alpha)$ we have

$$\Psi`\alpha = \sum_x(\alpha, \Psi`x) = \sum_x(\alpha, \Psi`x' - \Psi`x),$$

hence

$$\begin{aligned}
\aleph(\Psi`\alpha) &= \aleph(\textstyle\sum_x(\alpha, \Psi`x' - \Psi`x), \\
&= \aleph(\textstyle\sum_x(\alpha, \aleph(\Psi`x') \dot{-} \aleph(\Psi`x))) && \text{by [1], [2], 3.15, VII 3.5, 3.7;} \\
&= \aleph(\textstyle\sum_x(\alpha, \aleph(\Psi`x'))) && \text{by our induction premise;} \\
&= \aleph(\textstyle\sum_x(\alpha, \aleph(\Psi`x))) \\
&= \textstyle\sum_x(\alpha, \aleph(\Psi`x)),
\end{aligned}$$

since a sum of cardinals is a cardinal. Therefore if $\nu \in \aleph(\Psi'\alpha)$ then ν is in some $\aleph(\Psi'\beta)$ with $\beta \in \alpha$, so that our induction premise applies.

Thus on the whole we have

3.16 $$\{x \mid \mathrm{Od}(x)\} \ulcorner\!\Theta\!\urcorner\, \Pi.$$

From the construction of Θ it also follows that

3.17 $$a \in \Pi \;\&\; b \in \Pi \cdot\!\rightarrow\!\cdot\, \mathrm{dg}(a) \in \mathrm{dg}(b) \rightarrow \breve{\Theta}'a \in \breve{\Theta}'b$$

and hence by 3.13

3.18 $$a \in \Pi \;\&\; b \in \Pi \cdot\!\rightarrow\!\cdot\, a \in b \rightarrow \breve{\Theta}'a \in \breve{\Theta}'b.$$

Now with the aid of the Fundierungsaxiom we come to state that Π is the class V of all sets. In fact we have

$$\begin{aligned} c \cap \overline{\Pi} = 0 &\rightarrow c^* \subseteq \Pi \\ &\rightarrow c \;\in \Pi \qquad \text{by 3.14} \\ &\rightarrow c \;\notin \overline{\Pi}, \end{aligned}$$

hence

$$\overline{(Ex)}\,(x \in \overline{\Pi} \;\&\; x \cap \overline{\Pi} = 0).$$

Therefore by 2.5

$$a \notin \overline{\Pi},$$

thus

3.19 $$(x)(x \in \Pi)$$

and

3.20 $$\{x \mid \mathrm{Od}(x)\} \ulcorner\!\Theta\!\urcorner\, V.$$

With the help of the one-to-one correspondence Θ we can define $\sigma(A)$ in such a way that A'_σ, A''_σ and F'_σ become derivable. Namely defining

Df 3.5 $$\sigma(A) = \Theta'\mu_x(Ey)(y \in A \;\&\; x = \breve{\Theta}'y),$$

we can prove

3.21 a) $$a \in A \rightarrow \sigma(A) \in A$$

b) $$A \equiv B \rightarrow \sigma(A) = \sigma(B)$$

3.22 $$\sigma(A) \cap A = 0.$$

Indeed denoting the μ-term in Df 3.5 by $\mathfrak{n}$, we have

$$a \in A \rightarrow (Ey)(y \in A \;\&\; \mathfrak{n} = \breve{\Theta}`y)$$

[1] $$c \in A \rightarrow \mathfrak{n} \subseteq \breve{\Theta}`c.$$

The first formula gives

$$a \in A \rightarrow \Theta`\mathfrak{n} \in A$$

and thus 3.21 a) results. 3.21 b) is by Df 3.5 immediate. As to 3.22 we have

$$\begin{aligned} c \in \sigma(A) \cap A &\rightarrow c \in \Theta`\mathfrak{n} \;\&\; c \in A \\ &\rightarrow \breve{\Theta}`c \in \mathfrak{n} \;\&\; \mathfrak{n} \subseteq \breve{\Theta}`c \qquad \text{by 3.18 and [1]} \\ &\rightarrow \breve{\Theta}`c \in \breve{\Theta}`c; \end{aligned}$$

but $\breve{\Theta}`c \notin \breve{\Theta}`c$ since $\breve{\Theta}`c$ is an ordinal.

An immediate consequence of 3.20 is that every class A is in a one-to-one correspondence with a class of ordinals. Now by what we stated at the end of V, § 3, this class of ordinals is either represented or is in a one-to-one correspondence with $\{x \mid \mathrm{Od}(x)\}$. In the first case A, by II, 3.8, is also represented, and in the other case a one-to-one correspondence between A and $\{x \mid \mathrm{Od}(x)\}$ and also between A and V results. Thus we have the von Neumann statement, already intended by Cantor, that for every non represented class there is a one-to-one correspondence with $\{x \mid \mathrm{Od}(x)\}$, so that for every class there exists a wellorder. Formally we can express the said theorem, with explicitly indicating the one-to-one correspondence, by the formula (cf. IV, 3.3)

3.23 $$\mathrm{Rp}(A) \vee A \ulcorner_{\breve{\Theta}/\breve{A}\mathfrak{F}} \urcorner \{x \mid \mathrm{Od}(x)\},$$

$\mathfrak{F}$ being $\{xy \mid \mathrm{Sq}(x) \;\&\; \Delta_2 x^* \subseteq \mathfrak{K} \;\&\; y = \mu_z(z \in \mathfrak{K} \;\&\; z \notin \Delta_2 x)\}$

and $\mathfrak{K}$ being $\{z \mid (Eu)(u \in A \;\&\; \langle z, u\rangle \in \Theta)\}$.

We add here a consideration concerning the form of the axiom of choice used in the last reasonings of this section. A_σ was applied twice, once for the construction of the class Θ and then implicitly in the proof of 3.19 by the application of 2.5, which was reduced in the preceding section to A 7 by means of A_σ.

As we observed already, 2.5 can be derived from A 7 without applying an axiom of choice. This goes using the construction of Π, which essentially employs A 4. The derivation, due to Gödel, is as follows:

Firstly we prove anew $\Pi \equiv V$, but using A 7 instead of 2.5. By contraposition of 3.14 we get

[1] $$a \notin \Pi \rightarrow (Ex)(x \in a \; \& \; x \in \overline{\Pi});$$

by III, 4.8 we have

[2] $$(b \in a \rightarrow b \in \lceil a^* \rceil) \; \& \; \mathrm{Trans}\,(\lceil a^* \rceil)$$

and by [1] and [2]

[3] $$a \notin \Pi \rightarrow \lceil a^* \rceil \cap \overline{\Pi} \neq 0.$$

Further by Trans $(\lceil a^* \rceil)$ we get

[4] $$b \in d \; \& \; d \in \lceil a^* \rceil \cap \overline{\Pi} \rightarrow b \in \lceil a^* \rceil$$

and by once more applying 3.14

[5] $$d \in \lceil a^* \rceil \cap \overline{\Pi} \rightarrow (Ex)(x \in d \; \& \; x \in \overline{\Pi}).$$

The formulas [3], [4] and [5] together yield

[6] $$a \notin \Pi \cdot \rightarrow \cdot \lceil a^* \rceil \cap \overline{\Pi} \neq 0 \; \& \; (z)(z \in \lceil a^* \rceil \cap \overline{\Pi} \rightarrow \\ \rightarrow (Ex)(x \in z \cap (\lceil a^* \rceil \cap \overline{\Pi}))).$$

By VI, 5.21 and II, 2.4 the class $\lceil a^* \rceil \cap \overline{\Pi}$ is represented by a set. Application of A 7 to this set, together with [6], gives

[7] $$(x)(x \in \Pi) \text{ i.e. } \Pi \equiv V.$$

Now from $\Pi \equiv V$ it is easy to infer 2.5. Namely by the principle of least ordinal and [7] we get

$$a \in A \rightarrow (Ex)(x \in A \; \& \; (y)(y \in A \rightarrow \mathrm{dg}(x) \subseteq \mathrm{dg}(y)))$$

and by III, 1.11 and 3.13

$$\mathrm{dg}(b) \subseteq \mathrm{dg}(c) \rightarrow \mathrm{dg}(c) \notin \mathrm{dg}(b) \\ \rightarrow c \notin b.$$

Hence

$$a \in A \rightarrow (Ex)(x \in A \; \& \; (y)(y \in A \rightarrow y \notin x)),$$

what is the statement 2.5.

INDEX OF AUTHORS
to Part I

Ackermann, W., 5, 7
Archimedes, 4

Bachmann, H., 21, 24
Baer, R., 4, 24
Banach, S., 21
Bar Hillel, Y., 3, 15
Bernays, P., 6, 7, 9, 16, 18, 21, 25, 31–35
Bernstein, F., 28
Birkhoff, G., 20
Bolzano, B., 21
Borel, E., 20
Borgers, A., 4
Brouwer, L. E. J., 3

Cantor, G., 3, 8–11, 14, 16, 25–28, 31
Carnap, R., 23
Cavailles, J., 5
Church, A., 5, 16, 21, 23
Curry, H. B., 12

Dedekind, R., 21
Denjoy, A., 17, 20
Doss, R., 30

Essenin-Volpin, A. S., 18
Euclid, 3, 16

Finsler, P., 4
Firestone, C. D., 24
Foradori, E., 4
Fraenkel, A. A., 3, 5–7, 12–15, 17, 19, 21–23, 26, 30

Gandy, R. O., 17
Giorgi, G., 4
Gödel, K., 16, 25, 26, 31–34
Gonseth, F., 4, 17, 20

Hailperin, T., 6, 13
Hartogs, F., 20
Hausdorff, F., 20, 23, 24, 29
Hessenberg, G., 26
Hilbert, D., 4, 20, 23, 25

Kleene, S. C., 10
Knaster, B., 17
Kondo, M., 77
König, J., 28
Kuratowski, C., 9, 24, 26, 29

Lebesgue, H., 17, 20
Levi, B., 16, 17
Lindenbaum, A., 13, 19, 20, 22
Lorenzen, P., 4
Löwenheim, L., 34

McNaughton, R., 4, 34, 35
Mahlo, P., 23
Mendelson, E., 18
Merzbach, J., 4
Mirimanoff, D., 24
Montague, R., 35
Mostowski, A., 13, 18, 19, 30, 34

Neumann, J. von, 6, 7, 12, 13, 16, 22–25, 30–34
Novak, I. L., 34

Peano, G., 23

Quine, W., V., 4, 6, 7, 13, 25

Ramsey, F. P., 4, 16
Robinso(h)n, A., 6, 8
Robinson, R. M., 21, 32
Rosser, J. B., 11, 12, 21, 24, 34
Russell, B., 15, 16

Schoenflies, A., 4
Schröder, E., 12, 28
Shoenfield, I. R., 18, 34
Sierpiński, W., 20, 23
Skolem, T., 12, 13, 16, 22, 23, 34
Souslin, M., 18
Specker, E., 18, 20
Steinitz, E., 20
Suszko, R., 23
Szmielew, W., 19
Szpilrajn, E., 29

Tarski, A., 7, 18–24
Ting-Ho, T., 4

Wang, H., 4, 13, 21, 34, 35
Weyl, H., 5

Zermelo, E., 3–35 *passim*
Zorn, M., 20

INDEX OF SYMBOLS
to Part II

PREDICATES

		Page
Adp(S, F)	IV, Df 1.5	101
Adp(s, F)	IV, ,, 1.6	101
Adp(a, b, F)	IV, ,, 1.7	102
Alt(d)	III, ,, 1.2	80
Cd(m)	VI, ,, 3.3	141
Ch(D, C)	V, ,, 2.11	123
Ch(d, c)	V, ,, 2.12	123
Cl(a)	VI, ,, 4.1	144
Cls(c, a)	VII, ,, 5.3	190
Crs(K)	I, ,, 4.3	61
Crs(f)	II, ,, 4.3	76
Denum(a)	VI, ,, 5.2	152
Denum(A)	VI, ,, 5.3	152
Dsq(s)	VI, ,, 5.4	153
Fch(A) Fch(a)	VI, ,, 4.2	145
Fin(a)	III, ,, 5.1	97
Fin(A)	III, ,, 5.2	97
Ft(F)	I, ,, 4.1	61
Ft(f)	II, ,, 4.2	76
Ftp(d)	VII, ,, 1.1	156
Fund(d)	III, ,, 1.3	80
Gen(s, G)	IV, ,, 3.2	109
Gr(f, a)	VII, ,, 5.1	188
Gt(λ, α)	VII, ,, 2.3	170
Infin(a) Infin(A)	VI, ,, 5.1	148
It(s, a, F)	III, ,, 4.2	92
La(C)	V, ,, 2.13	124
La(c)	V, ,, 2.14	124
Lim(c)	III, ,, 2.3	88
Nc(A)	VI, Df 3.2	140
Nft(F)	IV, ,, 2.5	108
Nu(n)	III, ,, 3.1	89
Num(s, c)	IV, ,, 3.1	109
Od(d)	III, ,, 1.4	80
Or(C, A) Or(C, a)	V, ,, 2.1	118
Or(c, a)	V, ,, 2.2	119
Or(C)	V, ,, 2.3	119
Or(c)	V, ,, 2.4	119
OrS(D, a)	V, ,, 2.5	120
Po(C, A) Po(C, a)	V, ,, 2.9	123
Po(c, a) Po(C) Po(c)	V, ,, 2.10	123
Prog(F, C)	IV, ,, 1.4	101
Ps(A)	I, ,, 3.11	60
Ps(a)	II, ,, 4.1	76
Re(c)	VII, ,, 1.6	157
Rp(A, a)	I, ,, 4.5	63
Rp(A)	I, ,, 4.6	63
Sgr(b, f, a)	VII, ,, 5.2	189
Sq(s)	III, ,, 4.1	92
Sq(S)	IV, ,, 1.1	100
Sqa(s)	VII, ,, 2.1	167
Suc(c)	III, ,, 2.2	88
Trans(d)	III, ,, 1.1	80
Trans(D)	III, §4, Text	95
Wor(C, a)	V, Df 3.1	125
Wor(c, a)	V, ,, 3.3	126

		Page
Wor(c)	V, Df 3.4	126
WorS(C, a)	V, „ 3.2	125
$a \subseteq b$	I, „ 2.2	55
$a \subset b$	I, „ 2.3	55
$A \equiv B$	I, „ 3.1	57
$A \subseteq B$	I, „ 3.2	58
$A \sqcap_{K} B$	I, „ 4.4	62
$a \sim b$	II, „ 4.8	77
$a \preceq b$	V, Df 1.1	114
$a \prec b$	V, „ 1.2	114
$C, A \sqcap_{K} D, B$	V, „ 2.6	121
$c \approx d$	V, „ 2.7	122
$p =^{r} q$	VII, „ 1.3	156
$p <^{r} q$	VII, „ 1.4	156
$p >^{r} q$	VII, „ 1.5	156

FUNCTORS AND OPERATORS

		Page
$\iota_{\mathfrak{x}}\mathfrak{A}(\mathfrak{x})$	I, Df 2.1	54
$\bar{A}$	I, „ 3.3	58
$A \cup B$	I, „ 3.4	58
$A \cap B$	I, „ 3.5	58
V	I, „ 3.6	58
Λ	I, „ 3.7	58
$\bigcup A$	I, „ 3.8	59
$\bigcap A$	I, „ 3.9	59
$\{\mathfrak{x}\mathfrak{y} \mid \mathfrak{A}(\mathfrak{x}, \mathfrak{y})\}$	I, „ 3.10	59
$\Delta_1 A$	I, „ 3.12	60
$\Delta_2 A$	I, „ 3.13	60
$\breve{A}$	I, „ 3.14	60
$A \mid B$	I, „ 3.15	60
$A \times B$	I, „ 3.16	60
$F‘a$	I, „ 4.2	61
a^*	I, „ 4.7	63
$\overset{\prec}{A}$	I, „ 4.8	64
$[a]$	II, „ 1.1	66
$[a, b]$	II, „ 1.2	66
$\langle a, b \rangle$	II, „ 1.3	66
$\sum(m, F)$	II, „ 1.4	67
$\sum m$	II, „ 1.5	68
$a \cup b$	II, „ 1.6	68
$b \cap A$	II, „ 2.1	71
$b \cap c$	II, Df 2.2	71
$b^{-}c$	II, „ 2.3	71
$\bigcap_{\mathfrak{x}} \mathfrak{A}(\mathfrak{x})$	II, „ 2.4	72
$\prod_{\mathfrak{x}}(m, \mathfrak{t}(\mathfrak{x}))$	II, „ 3.1	74
$a \times b$	II, „ 3.2	75
$f‘a$	II, „ 4.4	76
$\Delta_1 a$	II, „ 4.5a	76
$\Delta_2 a$	II, „ 4.5b	76
$a \mid b$	II, „ 4.6	76
$\breve{a}$	II, „ 4.7	76
$A^{\underline{c}}$	II, „ 4.9	78
μA	III, „ 1.5	85
$\mu_{\mathfrak{x}}\mathfrak{A}(\mathfrak{x})$	III, „ 1.6	86
c'	III, „ 2.1	87
$\dot{\mu} A$	III, „ 3.2	91
$J(F, a)$	III, „ 4.3	92
$m+n$	III, „ 4.4	93
$m \cdot n$	III, „ 4.5	93
m^n	III, „ 4.6	93
$\lceil A \rceil$	III, „ 4.7	95
mlt(a)	III, „ 5.3	98
sg(S, n)	IV, „ 1.2	100
sg(s, n)	IV, „ 1.3	100

		Page
$\mathsf{A}F$	IV, Df 1.8	102
$\mathrm{I}(G, a)$	IV, ,, 2.1	105
$a+b$	IV, ,, 2.2	106
$a \cdot b$	IV, ,, 2.3	106
a^b	IV, ,, 3.4	106
$\mathrm{no}(a)$	V, ,, 2.8	122
$a^{\underline{b}}$	VI, ,, 1.1	130
$\Pi_{\mathfrak{x}}(m, \mathrm{t}(\mathfrak{x}))$	VI, ,, 1.2	131
$\aleph(a)$	VI, ,, 3.1	139
$o(b)$	VI, ,, 3.4	142
$p +' q$	VII, ,, 1.2	156
$p -' q$, $p \cdot' q$	VII, § 1 Satz 1.2	156
$\frac{a}{c}, -\frac{a}{c}$	VII, § 1 Text	157
$\left[\frac{k}{l}\right]$	VII, § 1 ,,	158
$[0]$, $[k]$	VII, § 1 ,,	158
$p \# q$	VII, Df 1.7	158
$-p$	VII, ,, 1.8	158
$p \ulcorner q$	VII, ,, 1.9	159
$\lvert p \rvert$	VII, ,, 1.10	159
$\lvert p, q \rvert$	VII, ,, 1.11	159
$p \square q$	VII, Df 1.12	159
$p \boldsymbol{\#} q$	VII, ,, 1.13	159
$rc(p)$	VII, ,, 1.14	160
$\frac{p}{q}$ (double bar)	VII, § 1 Text	160
$\mathsf{D}(F)$	VII, Df 2.2	170
$\mathsf{l}(\lambda)$	VII, ,, 2.4	171
$a \mathbin{\boldsymbol{+}} b$	VII, ,, 3.1	173
$a \mathbin{\boldsymbol{\times}} b$	VII, ,, 3.2	173
$\boldsymbol{a^{\underline{b}}}$	VII, ,, 3.3	173
$\sum_{\mathfrak{x}, m} \mathrm{t}(\mathfrak{x})$	VII, ,, 3.4	174
$\prod_{\mathfrak{x}, m} \mathrm{t}(\mathfrak{x})$	VII, ,, 3.5	174
$\aleph, \aleph_a$	VII, ,, 4.1	182
$\beth, \beth_a$	VII, ,, 4.2	183
$\Sigma s, \Pi s$	VII, § 5 Text	191
Ψ	VIII, Df 3.1	203
Π	VIII, ,, 3.2	204
$\mathrm{dg}(a)$	VIII, ,, 3.3	204
Θ	VIII, ,, 3.4	206
$\sigma(A)$	VIII, ,, 3.5	207

PRIMITIVE SYMBOLS

		Page
Propositional connectives: $\mathfrak{A} \,\&\, \mathfrak{B}$, $\overline{\mathfrak{A}}$, $\mathfrak{A} \vee \mathfrak{B}$, $\mathfrak{A} \rightarrow \mathfrak{B}$, $\mathfrak{A} \leftrightarrow \mathfrak{B}$	I, § 1	45
Quantifiers, Logical operators: $(\mathfrak{x})\mathfrak{A}(\mathfrak{x})$, $(E\mathfrak{x})\mathfrak{A}(\mathfrak{x})$, $\{\mathfrak{x} \mid \mathfrak{A}(\mathfrak{x})\}$, $\iota_{\mathfrak{x}}(\mathfrak{A}(\mathfrak{x}), a)$	I, § 1	46
Primitive predicators: $a=b$, $a \in b$, $a \in A$	I, § 1	48
Symbols introduced by axioms: 0, $b; c$, $\sum_{\mathfrak{x}}(m, \mathrm{t}(\mathfrak{x}))$,	II, § 1	65
$\pi(a)$	VI, § 1	130
ω	VI, § 5	148
γ	VII, § 1	162
$\sigma(a)$, $\sigma(A)$	VIII, § 1	196, 197

INDEX OF MATTERS
to Part II

absolute value 159
abstract algebra 188
abstract theories 188 ff
AC, axiom of continuum 162–163
adapted 101 ff
adaptor 102 ff
algebrait sum and product 192
algebraic theory of rings 190 ff
analysis 155 ff
Anzahldefinition *see* multitude
archimedean property of order 161
arithmetic functions of natural numbers 93, 106
——————— of ordinals 105 ff
— operations on cardinals 173 ff
———— on fraction triplets 156 ff
———— on real numbers 158 ff
ascending ordinal limit sequence 167, 172, 183, 185
Aussonderungstheorem 69
Auswahlmenge 134
Belegungsklasse 78
Bernstein–Schröder theorem 114 ff
Boolean algebra for sets 71
Boolean operations on classes 58–59
brackets 46–47, 106
Cantor's paradox 118
Cantor's theorem 117 ff
cardinal arithmetic 173 ff, 179 ff, 184 ff
cardinal number 139, 141
chain 123–124
chain, maximal 142 ff
characterised by finite elements 144 ff
choice, axiom of 43, 97, 111, 116, 133 ff, 151, 163
choice, axiom of, strengthened 195 ff, 208
Church schema 48, 52, 58, 91, 198
class formalism 56 ff
closed 144 (*see* Zorn's lemma)
closure operation 190
comparability 116, 138
continuous 108, 168
continuity of the system of real numbers 161 ff
continuum hypothesis 188
convergence 163
conversion schema 48
correspondence, one-to-one 61 ff, 77 ff
coupling to the left 64
critical points 109, 168 ff
Dedekind definition of infinity 97, 150
definition, explicit 49 ff, 55, 58, 61
—, recursive 91, 93
degree 204
denumerability 152 ff
denumeration 154
description 49, 54
difference of real numbers 159
distance 159
division (of ordinals) 165, 180
dots *see* brackets
Eigenschaft, definite 40, 41, 69
element relation 48, 65
equality 48, 52, 53, 57, 197, 198
equivalence theorem 114
Ersetzungsaxiom, *see* replacement, axiom of
extensionality axiom 52, 66.
finite 97, 147, 151
function 61
function, representation of, 73 ff
functional set 76
function theory (in analysis) 162

Fundierungsaxiom 200 ff
fundiert 80 ff
fraction triplet 155 ff
general recursion theorem, *see* recursion theorem
general set theory 65, 100, 130
—, weakened 89, 96, 99
generated by a sequence 170
generated numeration 109 ff
group theory 188 ff
Hausdorff principle 142 ff, 197
— recurrence formula 185
Hilbert ε-formula 197
Hilbert space 162
induction, numeral 90
induction, transfinite 86
infinity, axiom of 89, 109, 125, 147 ff
infinity 97, 148
initial number 140
initial section 120, 157
intersection 58, 71
interval 162
irreducible 172
iteration sequence 92
iteration theorem 92
iterator, numeral 92
iterator, transfinite 105
k-tuplet 67, 94, 107
König's theorem 177 ff
lattice 124
least natural number (principle of) 91
least ordinal number (principle of) 84
limit number 88, 166
limit sequence 167, 172, 183, 185
mapping class 62
measure theory 162
member of an ordered pair 59
member of a sequence 92
monomial 191
monotonic, *see* strictly monotonic
multiplicative axiom 134, 198
multitude 97 ff
mutually exclusive 58
n-complex 191
n-segement, *see* segment
natural number 89
natural order 122
negative fraction triplet 156
Normalfunktion 108 ff, 168 ff
numeration 109, 125 ff
—, generated 109 ff
numeration theorem 138 ff, 197
0 (symbol, constant) 54, 65
nulltriplet 156
onetriplet (1-triplet) 157
order 118 ff, 151
order, induced 121, 126
—, natural 122, 126 ff
—, partial 123, 142
—, restricted to a class 119
— type 122, 141 ff
ordinal (ordinal number) 80 ff, 201, 202
— arithmetic functions 105 ff, 164 ff
— function 105 ff
—, higher, lower 84
ω-section 166
ω-sequence 153, 202
pair class 60
pair set 76
pair, ordered 41, 59, 66
pair, unordered 66
parentheses, *see* brackets
Peano axioms 89, 90
polynomials 191 ff
positive fraction triplet 156
potency axiom 41, 130 ff
potency, cardinal 173
power 114 ff, 116, 137
predicate calculus 45 ff
primitive predicators 48
primitive recursion, *see* recursion
product, cardinal 173
Produktmenge (Cantor's) 131, 132

proper subset relation 55
rational (rational number) 155 ff, 158
real numbers 155, 157
recursion, ordinal 107
—, primitive 91, 93 ff
— theorem 102, 147
—, transfinite 100, 104 ff
reducible 172
reduction laws (in cardinal arithmetic) 186, 187
relation 59
replacement, axiom of 74, 100
— operator 74
—, theorem of 74
representations of classes by sets 63, 73, 130, 132, 148, 162
rest 165
ring theory 190 ff
rules of derivation 47
segment 100
sequence 92, 100, 153
sequential class 100
specialised variables 51, 55, 164
strictly monotonic 108, 167
subclass relation 58
subsequence 100
subset relation 55
subtraction of ordinals 165
successor 87, 88
sum axiom 41, 65
sum, cardinal 173
sum theorem 73
symbolische Auflösung 195, 196, 198
symmetrical polynomial 192, 194
Teichmüller principle 144 ff
term, class- 46
term, set- 46, 68
terminal section 120
topology of spaces 190
transfinite arithmetic 179 ff
transitive 80
transitive closure 95 ff, 154
triplet 67, 155 ff
type of a limit number 170 ff, 183–184
types, theory of 39, 42, 56
unit set 66
value of a function 61
wellorder 124 ff
wellorder of ordinals 80, 81, 84
wellorder theorem 111, 138 ff, 151, 163
Zahlenklasse 140 ff, 152, 182
Zorn's lemma 144 ff, 197

LIST OF AXIOMS
to Part II

			Page
E 1, *E* 2	equality and extensionality axioms	I, § 2	52
A 1, *A* 2, *A* 3	axioms of general set theory	II, § 1	65
A 4	potency axiom	VI, § 1	130
A 5	axiom of choice	VI, § 2	137
A 6	axiom of infinity	VI, § 5	148
AC	axiom of continuum	VII, § 1	162
A_σ, A'_σ A''_σ	stronger choice axioms	VIII, § 1	196, 197, 198
A 7, F_σ, F'_σ	axioms of Fundierung	VIII, § 2	201, 202, 203

BIBLIOGRAPHY
to Part I and II

Abbreviations: *Acad. USA* = Proceedings of the National Academy of Sc. (U.S.A.), *AMS* = American Math. Society, *FM* = Fundamenta Math., *J.* = Journal, *JSL* = Journal of Symbolic Logic

1937 W. ACKERMANN, Die Widerspruchsfreiheit der allgemeinen Mengenlehre. *Math. Annalen* **114**, 305–315.

1937a ———, Mengentheoretische Begründung der Logik. *Ibid.* **115**, 1–22.

1956 ———, Zur Axiomatik der Mengenlehre. *Ibid.* **131**, 336–345.

1955 H. BACHMANN, *Transfinite Zahlen.* Berlin–Göttingen–Heidelberg. 204 pp.

1956 ———, Stationen im Transfiniten. *Zeitschr. f. math. Logik u. Grundl. d. Math.* **2**, 107–116.

1928 R. BAER, Über ein Vollständigkeitsaxiom in der Mengenlehre. *Math. Zeitschr.* **27**, 536–543.

1929 ———, Zur Axiomatik der Kardinalzahlarithmetik. *Ibid.* **29**, 381–396.

1931 H. BEHMANN, Zu den Widersprüchen der Logik und der Mengenlehre. *Jahresb. d. D. Math.-Ver.* **40**, 37–48.

1937 P. BERNAYS, A system of axiomatic set theory. I. *JSL* **2**, 65–77.

1941 ———, *idem.* II. *Ibid.* **6**, 1–17.

1942 ———, *idem.* III. *Ibid.* **7**, 65–89.

1942a ———, *idem.* IV. *Ibid.* 133–145.

1943 ———, *idem.* V. *Ibid.* **8**, 89–106.

1948 ———, *idem.* VI. *Ibid.* **13**, 65–79.

1954 ———, *idem.* VII. *Ibid.* **19**, 81–96.

1948 G. BIRKHOFF, *Lattice theory.* Rev. ed. New York. 283 pp.

1920 B. BOLZANO, *Paradoxien des Unendlichen.* (Edited originally—posthumously—1851) *Philos. Bibl.*, No. 99. Leipzig. 157 pp.

1914 E. BOREL, *Leçons sur la théorie des fonctions.* 2me éd. Paris. 260 pp. (3me éd., 1928; 4me éd., 1950)

1949 A. BORGERS, Development of the notion of set and of the axioms for sets. *Synthese* **7**, 374–390.

1953 R. BÜCHI, Investigation of the equivalence of the axiom of choice and Zorn's lemma from the viewpoint of the hierarchy of types. *JSL* **18**, 125–135.

1947 R. CARNAP, *Meaning and necessity.* Chicago. 210 pp.

1954 ———, *Symbolische Logik.* Wien. 209 pp.

1938 J. CAVAILLÈS, Remarques sur la formation de la théorie abstraite des ensembles (Thèse). Paris. 156 pp.

1927 A. CHURCH, Alternatives to Zermelo's assumption. *Trans. AMS* 29, 178–208.

1932 ———, A set of postulates for the foundation of logic. *Ann. of Math.* (2) 33, 346–366.

1942 ———, Formal logic. *The Dictionary of Philos.*, ed. by D. D. Runes (New York), pp. 170–181.

1944 ———, *Introduction to mathematical logic.* I. (*Annals of Math. Studies*, No. 13.) Princeton N.J. 118 pp.

1934 H. B. CURRY, Functionality in combinatory logic. *Acad. USA* 20, 584–590.

1936 ———, First propeities of functionality in combinatory logic. *Tôhoku Math. J.* 41, 371–401.

1945 R. DOSS, Note on two theorems of Mostowski. *JSL* 10, 13–15.

1954 A. S. ESSENIN–VOLPIN, The unprovability of Suslin's hypothesis without the axiom of choice in the system of axioms of Bernays–Mostowski (Russian). *Doklady Akad. Nauk USSR*, N.S. 96, 9–12.

1926 P. FINSLER, Über die Grundlegung der Mengenlehre. *Math. Zeitschr.* 25, 683–713. (Cf. *ibid.*, 676–682.)

1933 ———, Die Existenz der Zahlenreihe und des Kontinuums. *Comment. Math. Helvet.* 5, 88–94. (Cf. J. J. BURKHARDT: *Jahresb. d. Deutschen Math.-Verein.* 48 [1938], 146–165, and 49 [1939], 146–155; S. LOCHER: *Comm. Math. Helv.* 10 [1938], 206–207.)

1949 C. D. FIRESTONE and J. B. ROSSER, The consistency of the hypothesis of accessibility (Abstract). *JSL* 14, p. 79.

1932 E. FORADORI, Grundbegriffe einer allgemeinen Teiltheorie. *Monatshefte d. Math. u. Phys.* 39, 439–454. (Cf. *ibid.* 40 [1933], 161–180; 41 [1934], 133–173; and the book *Grundgedanken der Teiltheorie*, Leipzig, 1937, 79 pp.)

1921/22 A. FRAENKEL, Zu den Grundlagen der Cantor-Zermeloschen Mengenlehre. *Math. Ann.* 86 (1922), 230–237. (Cf. *Jahresb. d. D. Math.-Ver.* 30 [1921], *97–98.*)

1922 ———, Über den Begriff "definit" und die Unabhängigkeit des Auswahlaxioms. *Sitzungsber. d. Preuss. Akad. d. Wiss.*, Ph.-Math. Kl., 1922, pp. 253–257.

1925–32 ———, Untersuchungen über die Grundlagen der Mengenlehre. I. *Math. Zeitschr.* 22 (1925), 250–273; II. *Journ. f. Math.* 155 (1926), 129–158; III. *ibid.* 167 (1932), 1–11.

1927 ———, *Zehn Vorlesungen über die Grundlegung der Mengenlehre.* Leipzig & Berlin. 182 pp.

1927a ———, Über die Gleichheitsbeziehung in der Mengenlehre. *J. f. Math.* 157, 79–81.

1928 A. Fraenkel, *Einleitung in die Mengenlehre*. 3. Aufl. Berlin. 424 pp. (Reprinted New York, 1946.) Cf. Fraenkel–Bar Hillel 1958.

1928a ———, Über die Ordnungsfähigkeit beliebiger Mengen. *Sitz. Preuss. Ak.*, Ph.-Math. Kl., 1928, pp. 90–91.

1937 ———, Über eine abgeschwächte Fassung des Auswahlaxioms. *JSL* 2, 1–25. (Cf. *C. R. de l'Acad. des Sc. Paris* 192 [1931], p. 1072.)

1958 A. A. Fraenkel & Y. Bar-Hillel, *Foundations of set theory*. Amsterdam.

1956 R. O. Gandy, On the axiom of extensionality. *JSL* 21, 36–48.

1941 G. Giorgi, Riflessioni sui fondamenti primi della teoria degli insiemi. *Pontif. Acad. Sc., Acta* 5, 35–40.

1938 K. Gödel, The consistency of the axiom of choice and of the generalized continuum-hypothesis. *Acad. USA* 24, 556–557. (Cf. *ibid.* 25 [1939], 220–224.)

1940 ———, *The consistency of the axiom of choice and of the generalized continuum-hypothesis with the axioms of set theory.* (*Annals of Math. Studies*, No. 3.) Princeton N.J. 66 pp. Revised ed. 1951.

1944 ———, Russell's mathematical logic. *The Philosophy of Bertrand Russell* (ed. by P. A. Schilpp, Evanston & Chicago), pp. 123–153.

1947 ———, What is Cantor's continuum problem? *Amer. Math. Monthly* 54, 515–525.

1933 F. Gonseth, Sur l'axiomatique de la théorie des ensembles et sur la logique des relations. *Comm. Math. Helv.* 5, 108–136. (Cf. *L'Ens. Math.* 31 [1933], 96–114.)

1936 ———, *Les mathématiques et la réalité*. Essai sur la méthode axiomatique. Paris. 386 pp.

1941 ———, *Les entretiens de Zurich* [1938]. Publiés par F. Gonseth. Zurich [1941] 209 pp. (See, in particular, the contributions of H. Lebesgue and W. Sierpiński.)

1944 T. Hailperin, A set of axioms for logic. *JSL* 9, 1–19.

1954 ———, Remarks on identity and description in first-order axiom systems. *Ibid.* 19, 14–20.

1915 F. Hartogs, Über das Problem der Wohlordnung. *Math. Annalen* 76, 438–443.

1904 F. Hausdorff, *Der Potenzbegriff in der Mengenlehre*. Jahrber. d. Deutsch. Math. Ver. 13, 569–571.

1914 ———, *Grundzüge der Mengenlehre*. Leipzig. 476 pp. Reprinted New York, 1949. — Later editions: *Mengenlehre*, Berlin & Leipzig, 1927, 285 pp.; new ed. 1935, reprinted New York, 1944.

1952 H. Hermes & H. Scholz, Mathematische Logik. *Enz. d. Math. Wiss.* Bd. I 1, Heft 1, Teil 1. Leipzig. 82 pp.

1906 G. Hessenberg, *Grundbegriffe der Mengenlehre*. Göttingen. 220 pp.

1923 D. Hilbert, Die logischen Grundlagen der Mathematik. *Math. Ann.* 88, 151–165.

1934–39 ——— & P. Bernays, *Grundlagen der Mathematik*. Bd. I (1934), Bd. II (1939). Berlin, 471 + 498 pp.

1933 S. C. Kleene, A theory of positive integers in formal logic. *Amer. J. of Math.* 57, 153–173, 219–244.

1952 ———, *Introduction to metamathematics*. Amsterdam–Groningen & New York–Toronto. 550 pp.

1937 M. Kondô, Sur l'hypothèse de M. B. Knaster dans la théorie des ensembles de points. *J. of the Fac. of Sc. Hokkaido Univ.*, Ser. I, 6, 1–20.

1921 C. Kuratowski, Sur la notion d'ordre dans la théorie des ensembles. *FM* 2, 161–171.

1925 ———, Sur l'état actuel de l'axiomatique de la théorie des ensembles. *Ann. de la Soc. Polon. de Math.* 3, 146–147.

1907 H. Lebesgue, Contributions à l'étude des correspondances de M. Zermelo. *Bull. de la Soc. Math. de France* 35, 202–212.

1902 B. Levi, Intorno alla teoria degli aggregati. *Istituto Lomb. di Sc. e Lett., Rendiconti* (2) 35, 863–868.

1923 ———, Sui procedimenti transfiniti. *Math. Annalen* 90, 164–173. (Cf. T. Viola: *Boll. dell' Unione Mat. Ital.* 10, 287–294, 1931, and 11, 74–78, 1932.)

1934 ———, La nozione di "dominio deduttivo" e la sua importanza in taluni argomenti relativi ai fondamenti dell' analisi. *FM* 23, 63–74. (Cf. *Publ. Inst. de Mat.* [*Rosario*] 2 [1940], 177–208.)

1938 A. Lindenbaum & A. Mostowski, Über die Unabhängigkeit des Auswahlaxioms und einiger seiner Folgerungen. *C. R. de la Soc. des Sc. et Lettres Varsovie*, Cl. III, 31, 27–32.

1926 ——— & A. Tarski, Communication sur les recherches de la théorie des ensembles. *Ibid.* 19, 299–330.

1936 ——— & ———, Über die Beschränktheit der Ausdrucksmittel deduktiver Theorien. *Ergebnisse eines Math. Kolloquiums* (Menger) 7, 15–22.

1955 P. Lorenzen, *Einführung in die operative Logik und Mathematik*. Berlin–Göttingen–Heidelberg. 298 pp.

1953 R. McNaughton, Some formal consistency proofs. *JSL* 18, 136–144.

1956 E. Mendelson, Some proofs of independence in axiomatic set theory. *JSL* 21, 291–303.

1956a ———, The independence of a weak axiom of choice. *Ibid.*, 350–366.

1925 J. Merzbach, *Bemerkungen zur Axiomatik der Mengenlehre.* (Inauguraldiss. Marburg a. d. L.) 39 pp.
1917 D. Mirimanoff, Les antinomies de Russell et de Burali-Forti et le problème fondamental de la théorie des ensembles. *L'Enseign. Math.* 19, 37–52.
1917/20 ———, Remarques sur la théorie des ensembles et les antinomies Cantoriennes. *Ibid.*, 209–217; 21 [1920], 29–52.
1956 R. Montague, Zermelo–Fraenkel set theory is not a finite extension of Zermelo set theory (Abstract). *Bull. AMS* 62, p. 260. (Cf. R. McNaughton, *Proc. AMS* 5, 505–509, 1954.)
1938 A. Mostowski, O niezależności definicji skończoności w systemie logiki (Thesis). *Dodatek Rocznika Polsk. Tow. Mat.* 11, 1–54. (Cf. *C. R. Soc. Sc. Lettres Varsovie*, Cl. III, 31 [1938], 13–20.)
1939 ———, Über die Unabhängigkeit des Wohlordnungssatzes vom Ordnungsprinzip. *FM* 32, 201–252.
1945 ———, Axiom of choice for finite sets. *Ibid.* 33, 137–168.
1948 ———, On the principle of dependent choices. *Ibid.* 35, 127–130.
1951 ———, Some impredicative definitions in the axiomatic set theory. *FM* 37, 111–124. (With correction, *ibid.* 38, p. 238. 1952.)
1953 ———, On models of axiomatic systems, *Ibid.* 39, 133–158. (Cf. R. Montague, *Bull. AMS* 61, 172f. 1955.)
1923 J. von Neumann, Zur Einführung der transfiniten Zahlen. *Acta Lit. ac Sc. Univ. ... (Szeged)*, Sectio Sc. Math. 1, 199–208.
1925 ———, Eine Axiomatisierung der Mengenlehre. *J. f. Math.* 154, 219–240. (Cf. 155 [1926], p. 128.)
1927 ———, Zur Hilbertschen Beweistheorie. *Math. Zeitschr.* 26, 1–46.
1928 ———, Die Axiomatisierung der Mengenlehre. *Ibid.* 27, 669–752.
1928a ———, Über die Definition durch transfinite Induktion und verwandte Fragen der allgemeinen Mengenlehre. *Math. Ann.* 99, 373–391. (Cf. *ibid.*, 392–393.)
1929 ———, Über eine Widerspruchsfreiheitsfrage in der axiomatischen Mengenlehre. *J. f. Math.* 160, 227–241.
1948/51 Ilse L. Novak (I. N. Gál), A construction for models of consistent systems. *FM* 37 (1950), 87–110. (Submitted in 1948 as a Doctoral Dissertation to Radcliffe Coll., Mass.)
1936 W. V. Quine, Set-theoretic foundations for logic. *JSL* 1, 45–57.
1937 ———, New foundations for mathematical logic. *Am. Math. Monthly* 44, 70–80.
1940 ———, *Mathematical logic.* New York. 348 pp. (2nd printing, with a corrigendum, Cambridge Mass., 1947. Revised ed., 1951.)
1941 ———, Element and number. *JSL* 6, 135–149.

1941a W. V. QUINE, Whitehead and the rise of modern logic. *The philosophy of A. N. Whitehead.* Evanston & Chicago, pp. 127–163.

1942 ———, On existence conditions for elements and classes. *JSL* 7, 157–159.

1953 ———, New foundations for mathematical logic. *From a logical point of view* (Cambridge, Mass.), pp. 80–101. [Slightly changed version of the 1937 edition, with *Supplementary Remarks.*]

1926 F. P. RAMSEY, The foundations of mathematics. *Proc. of the London Math. Soc.* (2) 25, 338–384.

1939 A. ROBINSO(H)N, On the independence of the axioms of definiteness (*Axiome der Bestimmtheit*). *JSL* 4, 69–72.

1937 R. M. ROBINSON, The theory of classes. A modification of von Neumann's system. *JSL* 2, 29–36. (Cf. Bernays' review, *ibid.*, p. 168.)

1953 J. B. ROSSER, *Logic for mathematicians.* New York. 540 pp.

1950 ——— & H. WANG, Non-standard models for formal logic. *JSL* 15, 113–129. (*Errata*, p. IV.)

1906 B. RUSSELL, On some difficulties in the theory of transfinite numbers and order types. *Proc. of the London Math. Soc.* (2) 4, 29–53.

1921 A. SCHOENFLIES, Zur Axiomatik der Mengenlehre. *Math. Ann.* 83, 173–200. (Cf. *ibid.* 72 [1912], 551–561, and 85 [1922], 60–64.)

1954 J. R. SHOENFIELD, A relative consistency proof. *JSL* 19, 21–28.

1955 ———, The independence of the axiom of choice (Abstract). *Ibid.* 20, p. 202.

1919 W. SIERPIŃSKI, L'axiome de M. Zermelo et son rôle dans la théorie des ensembles et l'analyse. *Bull. de l'Acad. des Sc. de Cracovie*, Cl. des Sc. Math. et Nat., Série A, 1918, pp. 97–152. (Cf. *FM* 2 [1921], 112–118.)

1921 ———, Une remarque sur la notion d'ordre. *FM* 2, 199–200.

1930 ——— & A. TARSKI, Sur une propriété caractéristique des nombres inaccessibles. *FM* 15, 292–300.

1922/3 T. SKOLEM, Einige Bemerkungen zur axiomatischen Begründung der Mengenlehre. *Wiss. Vorträge, 5. Kongr. der Skandin. Math., Helsingfors 1922* (1923), pp. 217–232.

1929 ———, Über einige Grundlagenfragen der Mathematik. *Skrifter utgit Norske Vid.-Ak. Oslo*, I, 1929, No. 4. 49 pp.

1930 ———, Einige Bemerkungen zu der Abhandlung von E. Zermelo „Über die Definitheit in der Axiomatik". *FM* 15, 337–341.

1953 E. SPECKER, The axiom of choice in Quine's New Foundations for mathematical logic. *Acad. USA.* 39, 972–975.

1954 ———, Verallgemeinerte Kontinuumshypothese und Auswahlaxiom. *Archiv der Math.* **5**, 332–337.

1957 ———, Zur Axiomatik der Mengenlehre (Fundierungs- und Auswahlaxiom). *Zeitschr. f. math. Logik u. Grundl. d. Math.* **3**, 173–210.

1909 E. Steinitz, Algebraische Theorie der Körper. *J. f. Math.* **137**, 167–309. — New ed. by R. Baer & H. Hasse. Berlin & Leipzig, 1930. 155 + 27 pp. notes.

1951 R. Suszko, Canonic axiomatic systems. *Studia Philos.* **4**, 301–330.

1947 W. Szmielew, On choices from finite sets. *FM* **34**, 75–80.

1930 E. Szpilrajn, Sur l'extension de l'ordre partiel. *FM* **16**, 386–389.

1925 A. Tarski, Sur les ensembles finis. *FM* **6**, 45–95. (Cf. *ibid.*, **30** [1938], 156–163.)

1925a ———, Quelques théorèmes sur les alephs. *Ibid.* **7**, 1–14.

1935 ———, Zur Grundlegung der Booleschen Algebra. I. *Ibid.* **24**, 177–198.

1935a ———, Einige methodologische Untersuchungen über die Definierbarkeit der Begriffe. *Erkenntnis* **5**, 80–100.

1938 ———, Über unerreichbare Kardinalzahlen. *FM* **30**, 68–89.

1939 ———, On well-ordered subsets of any set. *Ibid.* **32**, 176–183.

1948 ———, Axiomatic and algebraic aspects of two theorems on sums of cardinals. *Ibid.* **35**, 79–104.

1939 O. Teichmüller, Braucht der Algebraiker das Auswahlaxiom? *Deutsche Math.* **4**, 567–577.

1938 Tseng Ting-Ho, *La philosophie mathématique et la théorie des ensembles* (Thèse). Paris. 166 pp.

1949 Hao Wang, On Zermelo's and von Neumann's axioms for set theory. *Acad. USA* **35**, 150–155.

1950 ———, A formal system of logic. *JSL* **15**, 25–32.

1952 ———, The irreducibility of impredicative principles. *Math. Annalen* **125**, 56–66. (Cf. already *Acad. USA* **36**, 479–484, 1950; also R. McNaughton, *Proc. AMS* **5**, 505–509, 1954.)

1954 ———, The formalization of mathematics. *JSL* **19**, 241–266.

1955 ———, On denumerable bases of formal systems. *Math. Interpretations of Formal Systems* (Amsterdam), pp. 57–84.

1953 ——— & R. McNaughton, *Les systèmes axiomatiques de la théorie des ensembles*. Paris & Louvain. 55 pp.

1946 H. Weyl, Mathematics and logic. *Amer. Math. Monthly* **53**, 2–13. (Cf. *ibid.*, 208–214.)

1925/27 A. N. Whitehead & B. Russell, *Principia mathematica*. Sec. ed. Vol. I (1925), Vol II (1927), Vol III (1927). Cambridge, England. 674 + 742 + 491 pp.

1904 E. ZERMELO, Beweis, dass jede Menge wohlgeordnet werden kann. *Math. Ann.* 59, 514–516.
1908 ———, Neuer Beweis für die Möglichkeit einer Wohlordnung. *Ibid.* 65, 107–128.
1908a ———, Untersuchungen über die Grundlagen der Mengenlehre. I. *Ibid.*, pp. 261–281. (Has not been continued.)
1929 ———, Über den Begriff der Definitheit in der Axiomatik. *FM* 14, 339–344.
1930 ———, Über Grenzzahlen und Mengenbereiche. *Ibid.* 16, 29–47.
1935 ———, Grundlagen einer allgemeinen Theorie der mathematischen Satzsysteme. I. *Ibid.* 25, 136–146.
1935 M. ZORN, A remark on method in transfinite algebra. *Bull. AMS* 41, 667–670.

Additional bibliography to the second printing.

1965 A. ABIAN, *The Theory of Sets and Transfinite Arithmetic*. Philadelphia and London, 406 pp.
1967 H. BACHMANN, *Transfinite Zahlen*, 2. Aufl., Berlin-Heidelberg-New York, 228 pp.
1961 P. BERNAYS, Zur Frage der Unendlichkeitsschemata in der axiomatischen Mengenlehre, in "*Essays on the Foundations of Mathematics*" dedicated to A. Fraenkel, Jerusalem, pp. 1–49.
1963 P. J. COHEN, A minimal model for set theory, *Bull. Amer. Math. Soc.* 69, pp. 537–540.
1963 ———, The independence of the continuum hypothesis, I *Proc. Nat. Acad. Sc.* 50 (1963), pp. 1143–1148, II 51 (1964), pp. 105–110.
1961 P. ERDÖS & A. TARSKI, On some problems involving inaccessible cardinals, in "*Essays on the Foundations of Math.*" dedicated to A. Fraenkel, Jerusalem, pp. 50–82.
1963 P. FINSLER, Totalendliche Mengen, *Vierteljahrsschr. der Naturforschenden Gesellsch.* Zürich, Jahrg. 108, pp. 141–152.
1966 A. A. FRAENKEL, *Abstract set theory*, third ed., Amsterdam-North-Holland.
1961 G. B. KEENE, *Abstract Sets and Finite Ordinals*, 106 pp.
1963 A. J. KEISLER & A. TARSKI, From Accessible to unaccessible cardinals, *FM* 53, 225–308.
1964 D. KLAUA, *Allgemeine Mengenlehre*, Berlin, 581 pp.
1950 H. KNESER, Eine direkte Ableitung des Zornschen Lemmas aus dem Auswahlaxiom, *Math. Zeitschr.* 53, pp. 110–113.
1922 C. KURATOWSKI, Une méthode d'élimination des nombres transfinies des raisonnements mathématiques, *FM* 3, pp. 76–108.

1959 A. Levy, On Ackermann's set theory, *JSL* **24**, pp. 154–166.
1960 ———, Axiomschemata of strong infinity in axiomatic set theory, *Pacific Math.* **10**, pp. 223–238.
1960 ———, Principles of reflection in axiom set theory, *FM* 49, pp. 1–10.
1961 ———, Comparing the axioms of local and universal choice, in "*Essays on the Foundations of Mathematics*" dedicated to A. Fraenkel, pp. 83–90.
1961 ——— & R. Vaught, Principles of partial reflection in the set theories of Zermelo and Ackermann, *Pacific J. Math.* **10**, pp. 1045–1062.
1967 ——— & R. M. Solovay, Measurable cardinals and the continuum hypothesis, *Israel J. Math.* **5**, pp. 234–248.
1959 R. Montague & R. Vaught, Natural models of set theories, *FM* **47**, pp. 219–242.
1956 A. Mostowski, On models of Axiomatic Set Theory, *Bull. Acad. Pol. Sciences*, Vol. IV, pp. 663–667.
1965 ———, Thirty years of foundational studies, *Acta Philosophica Fennica*, Fasc. XVII, 180 pp.
1966 J. Schmidt, *Mengenlehre* I, Mannheim, 241 pp.
1951 J. C. Sheperdson, Inner models for set theory, I *JSL* **16** (1951) pp. 161–190, II *JSL* **17** (1952) pp. 225–237, III *JSL* **18** (1953) pp. 145–167.
1947 W. Sierpinski, L'hypothèse généralisée du continu et l'axiome du choix, *FM* **34**, pp. 1–5.
1960 P. Suppes, *Axiomatic set theory*, Princeton-New York-Toronto-London.
1954 G. Takeuti, Construction of the set theory from the theory of ordinal numbers, *J. Math. Soc. Japan* **6**, pp. 196–220.
1957 ———, On the theory of ordinal numbers, *Ibid.*, I Vol. **9** pp. 93–113, II Vol. 10 (1958), pp. 106–120.
1961 ———, Axioms of infinity of set theory, *Ibid.* Vol. **13**, pp. 220–233.
1961 ———, Remarks on Cantor's absolute, *Ibid.* **13**, pp. 197–206, part II *Proc. Jap. Acad.* **37**, pp. 437–439.
1954 A. Tarski, Theorems on the existence of successors of cardinals and the axiom of choice, Indagationes Math. **16**, pp. 26–32.
1955 E. J. Thiele, Ein axiomatisches System der Mengenlehre nach Zermelo und Fraenkel, *Zeitschr. f. math. Logik u. Grundl. d. Math.* **1**, pp. 173–195.
1950 H. Wang, The non-finitizability of impredicative principles, *Proc. Nat. Acad. Sc.* **36**, pp. 479–484.
1959 ———, Ordinal numbers and predicative set theory, *Zeitschr. f. math. Logik u. Grundl. d. Math.* **5**, pp. 216–239.

A CATALOG OF SELECTED

DOVER BOOKS

IN SCIENCE AND MATHEMATICS

CATALOG OF DOVER BOOKS

THE ELECTROMAGNETIC FIELD, Albert Shadowitz. Comprehensive undergraduate text covers basics of electric and magnetic fields, builds up to electromagnetic theory. Also related topics, including relativity. Over 900 problems. 768pp. 5⅜ × 8¼. 65660-8 Pa. $18.95

FOURIER SERIES, Georgi P. Tolstov. Translated by Richard A. Silverman. A valuable addition to the literature on the subject, moving clearly from subject to subject and theorem to theorem. 107 problems, answers. 336pp. 5⅜ × 8½. 63317-9 Pa. $9.95

THEORY OF ELECTROMAGNETIC WAVE PROPAGATION, Charles Herach Papas. Graduate-level study discusses the Maxwell field equations, radiation from wire antennas, the Doppler effect and more. xiii + 244pp. 5⅜ × 8½. 65678-0 Pa. $6.95

DISTRIBUTION THEORY AND TRANSFORM ANALYSIS: An Introduction to Generalized Functions, with Applications, A.H. Zemanian. Provides basics of distribution theory, describes generalized Fourier and Laplace transformations. Numerous problems. 384pp. 5⅜ × 8½. 65479-6 Pa. $11.95

THE PHYSICS OF WAVES, William C. Elmore and Mark A. Heald. Unique overview of classical wave theory. Acoustics, optics, electromagnetic radiation, more. Ideal as classroom text or for self-study. Problems. 477pp. 5⅜ × 8½. 64926-1 Pa. $12.95

CALCULUS OF VARIATIONS WITH APPLICATIONS, George M. Ewing. Applications-oriented introduction to variational theory develops insight and promotes understanding of specialized books, research papers. Suitable for advanced undergraduate/graduate students as primary, supplementary text. 352pp. 5⅜ × 8½. 64856-7 Pa. $9.95

A TREATISE ON ELECTRICITY AND MAGNETISM, James Clerk Maxwell. Important foundation work of modern physics. Brings to final form Maxwell's theory of electromagnetism and rigorously derives his general equations of field theory. 1,084pp. 5⅜ × 8½. 60636-8, 60637-6 Pa., Two-vol. set $23.90

AN INTRODUCTION TO THE CALCULUS OF VARIATIONS, Charles Fox. Graduate-level text covers variations of an integral, isoperimetrical problems, least action, special relativity, approximations, more. References. 279pp. 5⅜ × 8½. 65499-0 Pa. $8.95

HYDRODYNAMIC AND HYDROMAGNETIC STABILITY, S. Chandrasekhar. Lucid examination of the Rayleigh-Benard problem; clear coverage of the theory of instabilities causing convection. 704pp. 5⅜ × 8¼. 64071-X Pa. $14.95

CALCULUS OF VARIATIONS, Robert Weinstock. Basic introduction covering isoperimetric problems, theory of elasticity, quantum mechanics, electrostatics, etc. Exercises throughout. 326pp. 5⅜ × 8½. 63069-2 Pa. $8.95

DYNAMICS OF FLUIDS IN POROUS MEDIA, Jacob Bear. For advanced students of ground water hydrology, soil mechanics and physics, drainage and irrigation engineering and more. 335 illustrations. Exercises, with answers. 784pp. 6⅛ × 9¼. 65675-6 Pa. $19.95